Subramanian Senthilkannan Muthu
Editor

Detox Fashion

Case Studies

 Springer

Editor
Subramanian Senthilkannan Muthu
Bestseller
Kowloon
Hong Kong

ISSN 2197-9863   ISSN 2197-9871   (electronic)
Textile Science and Clothing Technology
ISBN 978-981-13-5229-4   ISBN 978-981-10-4783-1   (eBook)
DOI 10.1007/978-981-10-4783-1

This book was advertised with a copyright holder "The Editor(s) (if applicable) and The Author(s)" in error.

Printed on acid-free paper

This Springer imprint is published by Springer Nature
The registered company is Springer Nature Singapore Pte Ltd.
The registered company address is: 152 Beach Road, #21-01/04 Gateway East, Singapore 189721, Singapore

This book is dedicated to:
The lotus feet of my beloved
Lord Pazhaniandavar
My beloved late Father
My beloved Mother
My beloved Wife Karpagam
and Daughters—Anu and Karthika
My beloved Brother
Last but not least
To everyone working in the global textile
supply chain to make it TOXIC FREE &
SUSTAINABLE

Contents

Detoxifying the Supply Chains: Production Networks of Slow Garment Factories in South-Eastern Europe

WenYing Clarie Shih and Konstantinos Agrafiotis

Abstract An aspect of the detoxification processes of supply chains can be attributed to the reshoring of clothing production within the European Continent. Empirical evidence suggests that regional manufacturing networks operating in the South-Eastern fringes of Europe, after a devastating decade of diminishing orders, have recently started to produce specialized clothing for western European medium-sized niche fashion companies. These South-Eastern European manufacturers currently possess the experience in higher quality production. They also demonstrate some forms of sustainable production practices, and employ workers who are paid at least fair wages. Moreover, a certain sense of socio-cultural affinity emerges where all actors involved in the production and distribution acknowledge a loose interpretation of European solidarity which is translated into orders and subsequently into job retention within European borders. The theoretical underpinnings of this chapter lie in the concept of global production networks where commercial transactions among network members together with non-economic actors, such as civic associations and states can play an influential role in transnational production arrangements. The authors adopt the qualitative approach for this study based on the interpretivist methodology. In a single case study of a production network operating in the Balkans, the authors explore all major variables pertaining to the detoxification parameter of reshoring production. The findings broadly confirm that reshoring currently occurs in Europe as well as that all actors engage with sustainable practices which can form a viable and simultaneously competitive manufacturing strategy.

Keywords Nearshoring · Detoxification of the supply chains · Production networks · Sustainable production · Socio-cultural affinity

W.C. Shih (✉)
Department of Fashion Design, Hsuan Chuang University,
48, Hsuan Chuang Rd., Hsinchu 300, Taiwan
e-mail: wycshih@hcu.edu.tw

K. Agrafiotis
Independent Fashion Consultant, London, UK

© Springer Nature Singapore Pte Ltd. 2018
S.S. Muthu (ed.), *Detox Fashion*, Textile Science and Clothing Technology,
DOI 10.1007/978-981-10-4783-1_1

1

1   Introduction

The detoxification of clothing supply chains can take many forms as all these relate to sustainable development which currently constitutes a pressing issue in the fashion/clothing industry. An aspect of detoxification pertains to reshoring production back to the original locations of manufactures or near to the original locations.

Reshoring, a new term denoting that clothing production holds the potential of returning to European shores has gathered momentum in the recent past (Groom and Powley 2014). The European Parliament has commissioned research on the topic as it is particularly interested in remaining jobs in the sector and also exploring the possibilities of generating new jobs. The rationale for reshoring, in addition to the concerns regarding sustainable practices, can also be attributed to the so-called Total Landed Costs which have risen in other clothing manufacturing countries, such as China. This brings forth the notion of some production returning within European borders after decades of neglect (European Parliamentary Research Service 2014). Realistic benefits include the improvements in quality, certainty, production efficiencies and the made-in-Europe "label" with its heritage, quality, labor and environmental sensitivities that it connotes (Za 2014).

The theory of Global Production Networks (GPNs) helps to explain the reshoring flow of production in its European dimension. It is based on transnational relational networks which engage with production arrangements. Furthermore, the theory incorporates the so-called non-economic actors in the networks' configuration. For example, the role of the EU and its regional policies can induce and subsequently enhance the networks' performance and competitiveness. Other non-economic actors include civic associations which concern the environmental sustainability as well as the issue of workers' wages (Coe et al. 2008).

Due to the migration of production especially to China, peripheral Southeastern European networks after a devastating period seem to recover some production capacity as specialization, quality and proximity factors prevail in the reshoring logic (Pickles and Smith 2011). This can also be attributed to a loose interpretation of European solidarity as Western European quality brand owners have acknowledged the cachet of "made in Europe" production. This can also be attributed to mounting pressure induced by civic associations as customers want to know where their clothes are made and under what conditions (Friedman 2010).

In a single case study of a three partner company have recently formed a European cross-border production network. The authors explore the issues of reshoring and specialized quality production within European borders. Sustainable practices performed include slow fashion principles in the production processes, transparency in the supply chain, resource productivity in terms of economization of resources and machinery upgrading as well as labor conditions and wages. EU's regional policies are also addressed as they shape the manufacturing landscape and influence its competitive outcome. All actors involved in the case study namely, brand owners, the manufacturing company and EU's policies point to the direction that sustainable development features high on the production agenda as it forms a major concern in the detoxification of European supply chains in the Textile and Clothing sectors. The case also demonstrates that nearshoring more than reshoring to the original country of origin in this part of Europe. Nearshoring can be a viable strategy among manufacturers producing for upper-middle fashion brands as all parties involved are interested in keeping production within European borders, demonstrate sensitivities for the environment and also present convincing evidence of sustainable actions to their end customers. The thorny issue of living wages as opposed to fair wages is still debatable although the companies in the study have taken some corrective measures but still these are far from becoming reality for all garment workers in the region.

2 The Issue of Sustainability in the Fashion Industry

Sustainability in the fashion industry forms a fraction of the far bigger picture of sustainable development. However, it is widely recognized that the clothing industries including textile production are possibly the second most polluting industries after petrochemical companies worldwide (De Brito et al. 2008). Currently, the Earth's sustainable carrying capacity is at risk because of population growth, unethical business practices, rising affluence and patterns of wasteful consumption, all of which have prevailed across both developed and developing countries (Remy et al. 2016; Smitha 2011).

2.1 The Paris Agreement and Global Reawakening

The Paris Agreement on climate change, a global accord signed by 190 nations under the auspices of the United Nations, is historic as well as ambitious. Historic is because for the first time ever all UN members acknowledged the catastrophic human intervention on the planet. Ambitious is because the accord has set the bar high with regards to achieving the accord's targets. The Paris Agreement aims to decrease global warming to below 2 °C in the next decades bringing it to pre-industrial levels. The temperature target is set out of respect for developing countries which can be the most affected by global warming and rising sea levels (Donner 2015). The agreement also marks the beginning of the end of the global use of fossil fuels which can be served as the economic growth engine of the world's economy. Developed countries will mainly foot the bill of trillions of dollars in sustainable infrastructure programs around the world. For example, renewable energy sources including the use of solar and wind power as well as other renewable energy sources which will be invented and improved in the future (Donner 2015; Worland 2015). The deal is not exactly binding in all its articles but at least is the only one in existence which kept nearly all environmental activists and Non-Governmental Organization (NGO)'s satisfied with its long-term vision on emissions reduction aims and measures to safeguard transparency procedures (Worland 2015).

2.2 The Fast Fashion Mindset and Its Catastrophic Impact

In terms of the fashion industry, the fast fashion business model is certainly partly responsible for the global environmental deterioration (Cline 2013; Fletcher 2008). This model is based on shortening each season's time slots between production and retail by inducing customers to buy more as stores' shelves are replenished approximately every 30 days with fresh merchandise. Customers' attachment to this business model is to crave for new ranges every month because the traditional pattern of two fashion seasons is now replaced by approximately 6 fashion cycles of new merchandise per season. This is combined with the race to the bottom for the cheapest possible production location where fashion's flavor of the month has an irresistible price tag attached on it (Remy et al. 2016). In the event that one may add the recent deadly accidents in garment factories then the combination becomes certainly lethal in sustainability terms for both the environment and human cost (Cline 2013; Siegle 2011, 2016). This aggressive retailing model, praised by business analysts and economists alike, compels contract manufacturers to cut costs in every possible way (Remy et al. 2016; Siegle 2011; Smitha 2011). This is translated to the lowest common denominator in labor costs (especially women who are principally employed in garment factories), quality and environmental consid-erations since speed is one of the imperative concerns in the fashion retail sector (Barrientos 2013; Cline 2013).

3 The Fashion Industry Instigates Some Corrective Actions

Amid the dismal perspectives on sustainable development and human cost, both mainstream fashion and luxury have recently taken some actions to rectify the situation. Public awareness has played its part as an increasing number of customers who want to know where their clothes were made and under which conditions in terms of both environmental and labor. Ethical consumption can be refer to that consumers are not taking buying decisions solely on their personal interests but also incorporate societal and environmental parameters. However, it can be argued that in some cases, fashion companies which claim to be ethical, may engage in prac-tices which may not be entirely ethical with regards to the environment and/or workers' rights (Friedman 2010; Goworek 2011; Siegle 2016). These activities are termed *greenwashing* as they are usually devised by unscrupulous marketing and public relations managers who lead customers to believe that companies demonstrate deference to ethical practices including the environment (Siegle 2016).

Despite the dubious practices, a number of clothing companies in Europe together with trade bodies, universities, the European Parliament and the European Commission have also responded to sustainability demands by instigating public awareness, promoting ethical practices and forming associations (Goworek 2011). Civic society and non-governmental organizations, such as the Ethical Fashion Forum in the UK, the Clean Clothes Campaign in the Netherlands, and the European Parliament partner association, Fashion Revolution headquartered in the UK, pursue similar agendas by taking initiatives on the reduction of poverty of garment workers, cleaning up the supply chains, support sustainable practices and raise environmental standards in the textile and clothing sectors.

3.1 Transparency in Clothing Supply Chains

The traceability of materials and assembly operations in upstream production networks can form part of the sustainability architecture in the textiles and clothing (T&C) sectors (Germani et al. 2015; Martin 2013). The transformation of supply chains from the current state to a greener future can be a daunting task. Despite the difficulties, Lubin et al. (2010) argue that sustainability evolves into a mega business trend. The rationale is that companies wishing to achieve a competitive edge must view sustainability in its strategic context. They propose a road map based on a two pronged strategy which shall be pursued simultaneously: formulating a vision for value creation and taking actions. Companies need to redefine what they do in order to embrace this new source of value creation and also to transform systems and processes so that they can fit demands for sustainable practices. This transformation of systems and processes may lead to competitiveness (Germani et al. 2015).

The subject of the UN report on the traceability of supply chains may serve as a useful guide for the sustainability architecture in the T&C sectors. Traceability is defined as the ability to identify and trace the history, distribution, location and application of products, parts and materials (United Nations Global Compact Office 2014). In other words, it is a system monitoring the inputs as they enter the supply chain and are processed into end products. Incorporating the sustainability parameter into the system, traceability addresses also the sustainable origin of inputs, together with safeguarding good practices, respect for workers and obviously the environmental impact of these actions. Traceability is a similar concept to Life Cycle Analysis used in T&C sectors to help clothing companies to investigate the origin of their raw materials, and also administer improvements in the processes which are related to environmental issues (Seuring et al. 2008). To be noted, traceability is slowly becoming an accepted practice. In January 2014, the EU voted a set of directives on certification programs for companies operating within sustainable requirements. Coordination difficulties among stakeholders represent a

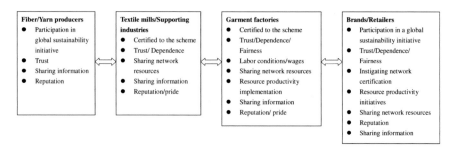

Fig. 1 A global collaborative scheme in the textile and clothing sectors—buyer driven global value chain *Adapted from* United Nations Global Compact Office 2014

considerable challenge as well as investments in technology remote manufacturing locations, language barriers and local managerial skills to access technology (United Nations Global Compact Office 2014). The traceability report recommends a best practice model based on the "Global Collaborative Scheme," which is shown in Fig. 1.

In the Global Collaborative Scheme, there must be coordinators who guide and work on traceability with all members in the given supply chain. Partners in the scheme participate according to their position in the chain and also communicate effectively about any problems with their immediate partners in the network (Curwen et al. 2013). To be noted, traceability is a difficult feat to achieve. Companies need to be aware that it is time consuming and must not be deterred by early disappointments. Partners in the network must establish interim evaluations of the progress and act according to their sustainability objectives. On the positive side, network members can enjoy reputational benefits since they have demonstrated a strong commitment to sustainability principles, once transparency has been achieved (United Nations Global Compact Office 2014).

3.2 Resource Productivity and Sustainable Materials Management

Another major issue concerning sustainable practices is how manufacturing companies deal with the waste in materials management practices and economization of resources. In manufacturing terminology, resource productivity (RP) refers to the sensible utilization of material resources in a production process by measuring the environmental impact. The goal is to use less and produce more with given resources. This entails cost efficiencies that contribute towards production optimization processes as economizing on resources saves money for clothing companies in a production network (Martin 2013). The Organization for Economic Cooperation and Development (OECD) (2011) subscribes to the same logic as the goal is to establish a resource efficient economy by enhancing resource productivity built on the 3 founding principles: *reuse-recycle-reduce* (*3R*). The report recommends that improvement in RP is achievable through sustainable materials management. This requires integrated life-cycle based strategies for waste, materials and products. Moreover, 3R initiatives are recommended such as circular rather than linear economy, integrated supply chain management, and finally encouraging the general public to take an ethically-based responsibility for sustainable growth.

RP in manufacturing can also be achieved from upgrading machinery, especially in garment factories where in many cases it is outdated. This does not represent an investment on a big scale although smaller manufacturers may find difficulties borrowing capital and buyers are usually indifferent to manufacturers' need for advanced machinery (Martin 2013). However, in the event of machinery upgrading, manufacturers calculate the business they can capture in accordance to the factory's capacity. This happens because the factory becomes more specialized so manufacturers may attract buyers of better brands thus participating into more upgraded networks. Advanced machinery can raise productivity in garment assembly, enhance quality of manufacturing (Abernathy et al. 1999; Shih and Agrafiotis 2015) and in turn, this can have a positive impact on workers' wages without necessarily triggering the next wave of relocation (Martin 2013).

3.3 *The Labor Conditions Issue*

It is a well-known fact that the T&C sectors are the first industries of any developing country along the path to industrialization (Dicken 2003). Especially the clothing industry is the most widely spread manufacturing sector and has long been the epicenter of the globalization process and a testing ground of workers' conditions (Cline 2013; Dicken 2003; Martin 2013). Unfortunately, the World Trade Organization (WTO) which is the official mediator of trade agreements has failed to include the social dimension in its provisions. This has led to a multiplicity of private initiatives addressing the issue of labor conditions (Bair et al. 2014).

Since the middle of 1990s, these initiatives have evolved into corporate social responsibility (CSR) programs which have led to auditing systems with the purpose of monitoring contractors' compliance. These programs are in essence risk management tools which can safeguard against reputational damage for brand owners/retailers, such as sweatshop scandals, which unfortunately occur often in garment factories (Bair et al. 2014; Cline 2013; Martin 2013).

Notwithstanding, CSR policies exercised by buying companies, the labor issue is far more complicated than a CSR program. It is not unusual that retailers/brand owners delegated inspections to independent auditors to be fraught with audit frauds. In this, auditors either turn a blind eye to breaches of labor and/or environmental regulations or recommend minor remedial actions (Siegle 2011). In other cases, fake certification documents are produced or buyers are shown demonstration factories but the majority of work is subcontracted to uncertified garment factories (Cline 2013; Martin 2013).

The problem is also compounded by the fact that least developed countries rely heavily on clothing exports, and simultaneously on the employment of large numbers of semi or unskilled workers. Moreover, in many countries, working conditions are often squalid and wages in garment factories are pitiful (Barrientos 2013; Cline 2013; Martin 2013). One of the major labor issues is workers' wages, especially living wages as factory owners may pay legal minimum wages but these are far from what the World Bank calculates as wages above poverty line (Barrientos 2013; Cline 2013). The cost pressure is constant to cut prices and in some cases, drastic cuts are demanded by buyers. It is sad that actual labor costs account for only 1% of the retail prices of fast fashion clothing articles, while sales are increasing, but clothing retail prices actually dropped 15% in the past decade (Cline 2013; Siegle 2016). In Eastern European countries, labor conditions are not a lot better. According to a recent report by the NGO Clean Clothes Campaign, it calls the "Made in Europe" label, the cheap labor garment factory for Western European fashion brands. Despite EU's pledge of reducing poverty and social exclusion by 2020, it seems that garment workers in the countries such as Bulgaria and Romania are still far from the living wages ideal. What is even more surprising in the report is that in 2013 legal minimum wages in these two countries were lower than China (Luginbühl and Musiolek 2014; Rankin 2014). Therefore, it is suggested in the literature (Barrientos 2013; Reimer 2009; Tokatli 2008) that it is naïve to believe that proximity to the Western EU markets and industrial upgrading can benefit garment workers, and women in particular, although some benefits arise for workers in more skilled positions such as pattern cutters and garment technicians.

4 The Slow Fashion Movement

Among the various initiatives for ethical fashion practices, the slow fashion movement is possibly one of the most prominent actions in recent time. The slow philosophy is promulgated by Honore (2004), and Fletcher (2008) in terms of clothing production and consumption. Both authors borrow heavily from the slow food movement instigated in Italy as a revolt to the penetration of fast food chains, although slow in this context is not literally meant to be the opposite of fast. Decelerating speed is about connecting to real and meaningful situations such as socializing with friends, participating in local culture, finding time to enjoy a meal, working to live and not living to work and so on. Slowness has more to do with a mental disposition this of being unhurried, reflective, calm, patient and valuing quality as opposed to quantity. Being slow does not mean regressing into a pre-Industrial Revolution utopia. On the contrary, it is more about finding a natural bio-rhythm, this of equilibrium where individuals can balance out the speed of contemporary life while simultaneously remaining slow inside (Fletcher 2008; Honore 2004).

Fletcher (2008) relates the slow philosophy to textiles and fashion by devising a number of fundamental precepts. In the slow fashion concept, the customer is conditioned to elect quality over quantity in the sense of consuming less and more responsibly. Companies within the slow fashion mindset engage themselves with fair practices towards their workers across the supply chain, thus drastically improving their livelihoods. They also reduce the use of raw materials by being more resourceful in alternative ways of preserving natural resources. Slow fashion firms use mostly local materials and labors instead of sourcing thousands of miles away from their base. In this, they can reduce carbon footprints. Moreover, they care about the preservation of local traditional skills. Aesthetic considerations are not consumed to the latest fad, but slow firms take a more neo-classic approach to design products with the quality of manufacture, which is certainly more durable. This guarantees that garments can be worn over longer periods of time. Financial viability is secured since slow companies can charge more for their clothing and accessories, and customers are willing to pay a premium because they know that fashion products are made in a fair practice ecosystem (Fletcher 2008).

In the next section of the chapter, the concept of reshoring is explained in its general sense as well as the specifics pertaining to the T&C sectors.

5   The Rationale of Production Reshoring

Reshoring describes a current situation where some production after decades of neglect may be returning to advanced economy countries. Reshoring as the name suggests is the opposite of *off-shoring* associated with the hollowing out of production in advanced economies, usually are with high worker salaries (Fernandes 2012; Groom and Powley 2014; Za 2014). According to the European Parliament Research Service (2014), reshoring is described as the partial or total return of production previously offshored to low-wage countries, to the original country, to serve local, regional or global demand. It is also referred to in-shoring, reverse offshoring, on-shoring, back-shoring and insourcing. After nearly three decades of shedding jobs in the manufacturing sector, manufacturers in the West off-shored production to low wage countries—where the major beneficiary was China—what is currently observed is a reverse offshoring movement (Wingard and Connerty 2014). However, this reshoring movement needs to be more closely examined as companies may return some production in the homeland, while simultaneously offshoring other parts of their production to cheaper locations. Moreover, reshoring production and job creation, despite the media reports, represent a tiny fraction of what they used to be in the years of mass manufacturing employment (Sirkin et al. 2011; The Economist 2013).

The rationale for reshoring production is not exactly crystal clear. Companies headquartered in the West have considerable operations in developing countries because these emerging economies contribute to an increasing share of retail sales and will continue to do so in the future. In other words, it seems to be a rational decision to continue manufacturing abroad (Sirkin et al. 2011; The Economist 2013; Wingard and Connerty 2014). Another important factor is the strategic and operational risks associated with reshoring. Companies need to examine their supply chains and see the network effects with regards to their component suppliers and cost escalation in seven crucial variables. These include: labor, staff shortages and training as well as higher wage structures since in Europe, labor cost is approximately 15 times higher than in China. Logistics/transportation, utilities, real estate, duties/taxes and incentives are provided by national governments and other regional government authorities. This amounts to the so-called *total landed cost*, which is a key measure for companies considering reshoring production as it calculates the end-to-end supply chain costs (Ellram et al. 2013; European Parliamentary Research Service 2014; Wingard and Connerty 2014).

Notwithstanding the risks, reshoring presents benefits in the following areas, these of improvements in the quality of finished products, protecting intellectual property, certainty and production efficiencies in the form of smaller batches, switching production schedules and introducing faster new products, decreasing transportation costs, and reducing disruptions in long supply chains. Moreover, reshoring is a desirable condition for national governments instigating social policies for the simple reason that reshoring creates jobs (Groom and Powley 2014; Wingard and Connerty 2014). It can be argued that for each manufacturing worker 2.5 additional jobs are generated in supporting industries. This is also supported by the fact that manufacturing R&D is higher than this in the service sector, with the additional R&D propensity for generating innovation and intellectual property (Hargreaves 2013). Lastly, domestic manufacturing holds the potential for exports increase and imports decrease (Groom and Powley 2014; EPRS 2014). Despite the perceived benefits of reshoring, a balanced sourcing strategy is recommended as companies need to trade off product categories with regards to the regions of manufacture, thus reaping comparative advantages (EPRS 2014; Groom and Powley 2014; Wingard and Connerty 2014).

6 Reshoring Production in the European Textiles and Clothing Sectors

Reshoring can be a realistic scenario since the Western European companies, especially under the Single Market Agreement, have sought to manufacture within EU borders, benefiting from the low labor wages of Eastern European countries combined with zero trade barriers such as customs duties and other protectionist impediments (Ellram et al. 2013; Van der Pols 2014). Furthermore, under EU's Outward Processing Trade arrangement with countries outside EU borders, partial duties are only paid on the added value of the assembled garments since fabrics and trims come from European mills (Cernat and Pajot 2012; Van der Pols 2014). The reshoring trend has been observed mainly among the Southern European countries of Italy, Spain, Portugal and Greece as well as Eastern European countries such as Bulgaria, Macedonia and Romania (Za 2014). Turkey remains one of the most favorable locations, not only for its proximity to the EU but for the fact that the country's textile mills, trim manufacturers and garment factories can provide complete end-to-end supply chain solutions which are appealing to retail brands. Simultaneously, Turkey's textile mills and garment manufacturers represent the biggest threat in terms of competitive pricing for European mills and garment manufacturers. This occurs because European retailers tend to be indifferent to manufacturing as usually their buyers seek the lowest possible prices (Van der Pols 2014). Despite Turkey's prominence, the "made in Europe" label is alluring in fashion. Since the Eurozone crisis has hit harder the economies of South European, states trade unions are willing to renegotiate wage terms with employers (Mellery-Pratt 2014; Za 2014).

This particular mode of production flows to South-Eastern Europe and Turkey can be better described as *nearshoring* because manufacturing does not in reality return to the original country of manufacture, but in nearer locations. Reshoring i can be observed in Western European countries such as Britain where manufacturers of luxury brands as well as upper-middle fashion companies have already shifted a part of their production to British shores. An increasing number of British consumers would like to purchase British-made fashion products. This is also due to the fact that consumers in emerging economies such as China prefer to buy British-made clothing and other fashion accessories (Hargreaves 2014). British designers are also prompted to use local fabrics and trims, and make clothing in Britain. However, after years of textile mills and clothing manufacture decline, there is a serious shortage of skillful workers. Italy is another example but it has managed to retain some production because the "made in Italy" label connotes superior craftsmanship and quality (Za 2014). Portugal also features highly on the near-shoring agenda as the country has managed remarkably to upgrade especially its leather and footwear industry and currently manufactures for famous Italian and French luxury and upper-middle brands as well as local brands with an increasingly international outlook. Frequent changes in design, smaller batches, quality improvements in materials and craftsmanship and timely deliveries to retail outlets constitute the reason of reshoring to those stated in the generic rationale of returning production to the homeland (Mellery-Pratt 2015).

In order to understand how transnational supply chain players function in relational networks in terms of textiles and clothing manufacturing, the authors in the following section discuss the theoretical underpinnings, namely the global production networks and their influence in commercial transactions and other non-economic factors such as the role of state and social parameters.

7 Global Production Networks

Globalization ushered the world in a new era of competition as production was massively outsourced to developing countries thus reconfiguring trade patterns and production on a global scale. This global shift has also resulted in the restructuring of companies to configure new formations in order to organize their activities

(Dicken 2003; Gereffi and Lee 2012). Reshoring can be analyzed using essentially two complementary theories these of the global value chains (GVCs) and global production networks (GPNs) (Fernandes 2012). The former is based on the transaction cost theory where the fundamental problem companies have to resolve with regards to their production is whether to internalize (vertical integration), or outsource production into a gradation of commercial engagements ranging from joint ventures and partnerships to spot markets (Gereffi et al. 2005). In the transaction cost theory, there are risks involved as outsourcing production, especially far from the home base, can be fraught with difficulties and uncertainty, thus companies need to hedge against uncertainty by forming detailed contracts (Fernandes 2012). The resource-based theory and more particularly the relational view espoused by the GPN theory, advocates that resources which reside within the firm can be combined with the resources deriving from external collaborations with another firm operating in a network (Dyer and Singh 1998; Lavie 2006). In this view, interdependencies are fundamental as they have to be ordered by agreements in a formal and/or informal manner (Coe et al. 2008). Nevertheless, trust is the prevailing concept as it reduces the need for contracts (Gereffi and Lee 2012).

More analytically, the concept of global value chains (GVCs) emerged in the 1990s as mainly Western buyers sought to outsource components, subassemblies and also finished goods from developing countries and especially from the Far East in a sequential transformation mode. GVCs encompass all linear activities that create value from design and materials supply to production and distribution. GVCs are essentially governance mechanisms (Gereffi and Lee 2012), and broadly can be divided into production-driven or buyer-driven. Buyer-driven chains refer mainly to textiles and clothing whereby the relational governance form corresponds to this study pertaining to European clothing manufacturing (Gereffi, Humphrey and Sturgeon 2005). Relational governance involves buyers and suppliers in complex decisions which cannot be easily codified. Thus, information combined with knowledge sharing and constant interactions among members in a network are essential. This is coupled with mutual trust and social rapports which are also critical in coordinating relational networks (Gereffi and Lee 2012).

The so-called Manchester School of Geographers expanded upon the concept of GVCs by offering a more comprehensive view regarding their structure (Coe et al. 2008; Dicken 2003). Global production networks (GPNs) can explain in a more cohesive manner how the global economy functions as the transnational network is critical in its geographical complexity for it reflects the relational and also the structural nature of how production, distribution and consumption is organized. Furthermore, GPNs exist in the transnational space where actors engage and shape the geographies of political, cultural and social conditions as well as the transformation process of production. This means that GPNs are not confined only to the "free market" relationship of buyers and suppliers advocated by GVCs, but to extend their scope by integrating other influencing actors into the equation. This occurs because market economy is socially and commercially constructed. Therefore, social forces and the state can play an influential role in this market configuration (Barrientos 2013; Coe et al. 2008). GPNs possess two fundamental differences with GVCs as they do not refer to linear flows and sequential stages but in complex circuit configurations. Additionally, GPNs integrate all sets of mediators and links involved in a *profound relationality mode* within the production, distribution, consumption network. The net-chain, according to Coe et al. (2008) is the prevailing theoretical framework which is comprised of a set of networks in horizontal and vertical configurations. Horizontal connections refer to relations among actors on the same layer of transactions, while vertical connections refer to commercial relations among firms on different layers. These horizontal and vertical net-chains can be either short or long depending on the number of companies involved in the vertical configuration (production/distribution sequences). The horizontal actors involve states, trade agreements, civic society associations, environmental factors, and labor issues. The interdependent nature of networks is of fundamental importance because it impacts governance and power factors within the network (see Fig. 2).

Power asymmetries in GPNs as in bargaining situations do exist but these are configured in a different manner from GVCs because it is not always the bigger player who wields power to the weaker one. Companies struggle for resources and positioning in the network against weaker commodity competitors. Nevertheless, companies are not static as they can evolve overtime depending on contingent circumstances prevailing in the network. This mobility may facilitate upgrading to a more elevated status in a network. In many cases, companies operating in a network are simultaneously the members of another network with upgrading opportunities

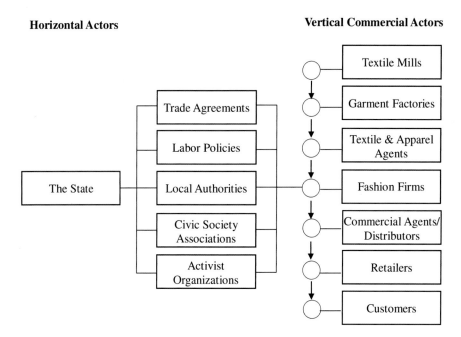

Fig. 2 The Net-Chain configuration of global production networks *Source* The Authors

(Coe et al. 2008). To be noted, GPNs do not solely engage with commercial parameters embedded in firms. They are also driven horizontally by non-economic agents where at times either all or parts of them are involved (Barrientos 2013). The role of states, regional powers such as the EU and geographic regions, civic associations, the World Trade Organization and various trade agreements can influence GPNs in varying degrees. These relationships can be cordial and cooperative, and simultaneously they can also be conflictual and competitive (Barrientos 2013; Coe et al. 2008).

8 Peripheral Production Networks in the European Continent

Europe's T&C manufacturing landscape has changed dramatically as Eastern European countries following deindustrialization after the collapse of the Soviet bloc in 1989 were re-inserted rapidly into European production networks, mainly serving for western European brand owners and branded manufacturers. The EU and its trade policies and customs agreements and the so-called Outward Processing Trade have had a profound leverage in manufacturing and trade in this region. However, after the abolition of the Multi-Fiber Arrangement (MFA) in 2005, and the subsequent wholesale relocation of production to mainly China and India, orders in Eastern European countries (EEC) dropped dramatically, thousands of jobs were lost, and factories closed, a fact that brought about renewed desperation and poverty in the region (Pickles and Smith 2011; Pickles et al. 2006).

Manufacturers' response came in a variety of ways. Some managed to upgrade their production capabilities by manufacturing garments of higher complexity such as tailoring which demanded better skills and commanded better prices especially in the case that design elements, fabrics and trim were provided by the manufacturers. Furthermore, following relocation eastwards, upgraded manufacturers started sub-contracting assembly operations in other Eastern countries such as Ukraine. Others directed their manufacturing capacity in producing low end garments for local markets, while others opted for full upgrading to brand owner status selling mainly to domestic markets following the emergence of local middle classes. Manufacturers who have had connections with buyers in Western Europe have managed to sell full package services to Western European brand owners. It must also be noted, that backward linkages, namely fabric and trim suppliers after years of experience with Western European buying firms, have had reached already a level of export quality. This was also triggered by mainly Italian and Greek textile mills which have had relocated part if not all of their production capacity to EEC countries. As a result, they trained local textile engineers in the production of superior materials (Pickles and Smith 2011; Pickles et al. 2006).

Following manufacturing reforms, EEC countries response to the so-called China price seems to have sustained some production as regions and countries have managed the specialization of their production by serving niche brand owners.

According to Pickles and Smith (2011) there are principally three reasons for this resilience in manufacturing. It can be attributed to the fact that Chinese price competition focuses on usually lower-end garment production where EEC manufacturers are unable to compete, thus, they have relinquished production in this clothing section and opted to upgrade to higher complexity garments. Proximity to the markets especially in the fast fashion sector is another important factor as retail replenishment needs to be rapidly manufactured and distributed to Western European retailers. Lastly, the quality and specialization factors come to the fore especially in tailored garments where fabrics, trims and assembly operations need to reach certain quality standards.

9 Research Methodology and Methods

The authors adopt the philosophical view of interpretivism in the logic of interpretation and observation in order to understand the context of the world. This approach is integral to the qualitative research, which is focused on the contexts of how humans live and work, which helps researchers to understand the cultural and historical background of companies and people in the specific situations. Moreover, the researchers' own setting can configure their interpretation because they position themselves in the situational investigation (Creswell 2009). Both authors' working experience in the clothing industry as well as the second author's experience as a consultant in the clothing sector for companies seeking alliance partners in the Black Sea the Balkans regions was very helpful. It has enabled them to comprehend and analyze in depth the subject matter. That is, the researchers' intent is to interpret a pattern of meanings which is configured by the actors who shape the research content.

Qualitative research methods are employed in this research because the study aims to explain subjective processes and meanings of how individuals or groups take action or cope in a particular situation based on the social construction of reality (Creswell 2009). Qualitative methods are deemed more appropriate for this study to understand the inherent meaning of reshoring production and sociability of ethical manufacturing in slow factories. The case study approach is employed in this study as it explicitly explores managerial thinking, perceptions, activities and expanding horizons with regards to current strategic moves in the Balkan region (Yin 2009). This may provide an in-depth understanding of how owners/managers perceive the transnational network of companies and how they operate in this part of the world, ranging from sustainable practices to garment factory workers livelihoods.

Interviews, a qualitative method, are used as the primary data collection method to represent rich, in-depth and relevant information for the study. Interviews allow researchers to explore and capture the insights of the research through the interviewees' explanations (Yin 2009). Unscripted questions with a flexible approach are also used in follow-up interview sessions, allowing interviewers to capture relevant details within a situation comfortable for the interviewees (Oppenheim 1992).

Semi-structured in-depth interviews with top managers who are also the owners of the textile/clothing company (Continental Jeans & Casuals) were conducted in order to understand their pertinent manufacturing perceptions in three country operations. Selection criteria included the fact that Continental was recently established in 2012 and as a holding company among three partners, each with over 30 year experience in the textile/clothing sectors. One of the partners after nearly a ten year residence as a textile development and clothing manufacture consultant in China returned to Europe to pursue sustainable manufacturing practices in jeans wear. He convinced the other two partners of his conviction to sustainable small batch manufacturing for niche fashion brands which demand superior fabrics and craftsmanship in their collections. The three partners joined forces in order to establish a complete sustainable model of manufacturing in slow factories thus, catering to the needs of highly specialized brand owners operating in European markets. The second major selection criterion is the fact the authors were particularly interested to explore how a predominately jeans and casuals company operates in relation to the well-known fact in the industry that jeans washing consumes vast amounts of water which can be a major polluter. The third criterion is associated with the company's garment factories. This study is to explore how the notion of slow manufacturing works, as well as how garment workers are treated in terms of wages and working conditions. In this regard, the authors wanted to observe firsthand how the company which claimed to use sustainable practices was behaving in reality.

The interviews were conducted during the summer (June–July) of 2016 in Thessaloniki, Greece and Petrich in Bulgaria. All interviews were transcribed and then sent back to the interviewees for clarification and verification as this can enhance the reliability of this study. Data analysis, according to Yin (2009) includes data examination, categorization and tabulation. It must be noted that qualitative data analysis relies heavily on the researchers' knowledge and understanding to interpret, analyze and categorize the data in order to make logical assumptions about the corresponding linkages between the collected data and its interpretation into the study's topic. Nevertheless, in order to avoid bias, the researchers have triangulated data not only from interviews, but also have made use of their personal observations and secondary sources such as trade documents and current affairs articles in local and international newspapers. In the next section, the authors discuss in detail the case study and its production arrangements in three European countries.

10 Case Study

The holding company, Continental Jeans & Casuals was established by three partners in 2012 as a joint venture incorporating textile, clothing and laundering facilities in Greece, Bulgaria and Romania. The initial two partners had twofold textiles and clothing operations in both knitted and woven materials. The first section of the operations, located in Romania, produced knitted cotton/cotton blend fabrics in circular knitting machines combined with the manufacture of t-shirts and sweatshirts for fast fashion European retail brands. In other section of the operations, woven fabric was bought from Greek and Turkish mills and made into jeans and casuals in the company's garment factories in Bulgaria. Jeanswear manufacture was combined with laundering and finishing at the company's facility in Northern Greece, also catering to fast fashion retail brands. The third partner who was residing in China as a chief T&C production consultant for big American and European corporations realized that within the calculations of the *total landed costs*, it was feasible for upper middle brands of jeans and casuals to return production in Europe. Moreover, he was fully aware that niche brands would demand superior quality materials, craftsmanship and sustainable practices. Prior to his arrival to Greece, he convinced the other two partners of shifting their attention to slow manufacturing by targeting niche European brand owners. To be noted that, at the time of the third partner's arrival and subsequent agreements with the other two, operations of the mills and factories in all three countries were running at less than 50% of their capacity as a significant part of orders per season had already shifted mainly to Turkey.

This forced downsizing was also responsible for nearly half of workers dismissals. Turnaround of the company's operations was urgently needed. The three partners proceeded in the formation of the holding company (Continental Jeans & Casuals), pooled financial resources together and embarked on participations to European clothing trade fairs in pursuit for niche brand owners. The other two partners realized the reshoring potential as part of sustainable manufacturing. They contacted like-minded brand owners who were looking for manufacturers procuring materials and making garments within Europe in smaller production batches with a sustainable manner. During the first year of trade fair participations, some brand owners placed initial test orders for the first season and these were converted into full orders the following season/year. This meant that the partners had to restructure rapidly the mills and the factories in order to produce smaller quantities of textile and clothing in superior quality and in a sustainable manner. Some new garment machinery for the Bulgarian and Romanian plants was bought and two old generation knitting machines were replaced while others were updated. Moreover, as the laundering plant located in Greece, an upgraded water treatment facility was installed complying with EU standards. The company then acquired all appropriate certification following technical audits. The depreciation of all new and upgraded machinery was planned to be amortized within seven years. All investments were deemed necessary in order to comply with EU regulations as well as with brand owners' requirements for sustainable production. Moreover, sustainable laundering and finishing techniques were introduced such as "dry washing" which consumes considerably less water (see Fig. 3).

The garment factories were in a good working condition which complied with EU's regulations regarding health and safety. Nevertheless, garment production needed to reform by hiring two young production engineers who understood slow production and sustainable practices. The engineers were responsible for supervising production in both knitted manufacture facilities in Romania and woven in

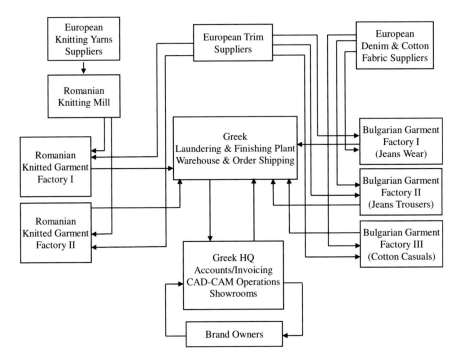

Fig. 3 The Mapping of Continental's operations

the Bulgaria facilities. Some of the laid-off garment workers were rehired as slower production required superior craftsmanship and more manual work. Additionally, all garment workers were retrained in slower production methods of smaller batches and more frequent style changes. Speed no longer constituted the imperative in production and instead both production engineers stressed the point of quality, craftsmanship and transparency in the supply chain. This retraining activity was partially subsidized and supervised by the EU's Regional Development Fund which advocated sustainable production practices. As far as fair wages are concerned, the company had and still has a decent record in its employment history with the local labor inspectorates in all three countries.

Deliveries of all initial test orders placed by the brand owners during the first season (Autumn/Winter 2013) were a success in terms of quality and timing. In the following three years, orders started to flow steadily to all of the company's facilities as consistency and quality of work, prompt deliveries and proved sustainable practices spurred other brand owners to place orders. During the time of interviews and the visits of factories, Continental had managed to pull out of its uncertain future, have a full order ledger and enjoy reputational benefits as a reliable quality European manufacturer which in turn has enhanced the company's competitiveness.

The following discussion based on four fundamental constructs identified in the literature include the traceability of the T&C supply chain, resource productivity, labor conditions and slow fashion. The four constructs are viewed from the perspective of the transnational production networks theory. To be noted that all constructs are not mutually exclusive as they are closely interrelated.

11   Analysis and Discussion

The literature regarding transnational production networks broadly concurs with the case study (Coe et al. 2008; Pickles and Smith 2011). However, European brand owners (buyers) were not dominant players in the supply chain. This occurred because they realized that they could not proceed with small batch production if they played manufacturers against each other on price and other corner-cutting unethical practices evident in other parts of the world (Cline 2013; Gereffi and Lee 2012). Thus, relational operations prevail. According to GPNs theory, the market economy is socially as well as commercially constructed so in this production configuration social forces can play an influential role (Coe et al. 2008). This was made clear to the authors during the interviews with Continental's owners/directors. They stated that a spirit of cooperation prevailed in their business transactions with the brand owners. Both parties realized that this is the only way forward. They refrained from power struggles as both understood their position in the supply chain. This partially contradicts the literature which advocates that competition for resources and positioning in the supply chain features high on power struggles among brand owners-manufacturers (Coe et al. 2008; Gereffi and Lee 2012). To be clear, power asymmetries do exist in these commercial relationships as Continental is a small company in comparison to some of its customers (brand owners) but this does not affect cooperation between them. Moreover, as revealed in the interviews, the state namely the EU did play an influential role as the upgrading of the water treatment machinery in the Greek laundry plant was partially subsidized by EU's Regional Coherence Funds. Water treatment falls within the parameters stipulated by the EU on sustainable development (European Commission 2015). Additionally, the EU's Regional Development Funds supported retraining programs for textile/garment workers in all three countries. Support came in the form of financial benefits to workers for a year and Continental had to hire consultants/instructors to train all workers involved in the program. This confirms the literature which supports the net-chain argument as commercial actors and the state are intertwined in transnational production configurations (Coe et al. 2008).

12   Nearshoring Production Relating to the Traceability of the T&C Supply Chain

Traceability features high on the agenda of brand owners as they are currently under pressure from activist organizations and the public to demonstrate ethical production practices in both environmental and labor issues. More specifically, it is a well-known fact in the industry that jeans manufacture worldwide has established a notorious reputation of polluting the environment due to laundering and finishing processes where toxic chemicals are used. Thus, the provenance of jeans production needs to be closely monitored (Greenpeace International 2012). This has repercussions for upper middle retailers and brand owners as their reputation is at stake by not conforming to transparent practices. Thus, nearshoring production features as a high priority in their sourcing decisions and Europe represents one of the best production locations despite higher retail prices. On the consumption side, European customers are willing to pay more for clothing staples such as jeans wear which can provide superior quality and durability along with the ethical guarantee. This has been revealed during the interviews as for Continental, it was not in reality a matter of compliance with EU regulations or pressure from activists or brand owners, but it was rather imperative to instigate these necessary changes because the viability of the business depended upon these changes. Moreover, Continental's owners stated that operating within Europe meant that companies on either side of the production/consumption equation needed to cooperate closely because markets, brands, products and manufacture are common in all European markets. There was no point of being unethical as this could easily backlash with detrimental effects for all businesses involved (manufacturers-brand owners-retailers). In other words, with regards to sustainable practices, they could see that jeans production was heading towards this ethical direction as it was either restructuring or be shamed and eventually cease operations. The above partially contradicts the literature which postulates that unethical practices prevail in the industry (Cline 2013; Siegle 2011) as in the case study, the supplying company (Continental) although operating in a fast fashion ecosystem realized that the future lied in ethical production and within a brief period of time switched completely to a fresh mindset and new slower production methods.

13 Resource Productivity Stands for Quality Rather Than Quantity and Speed

Restructuring and reforming the business model from fast fashion to slow/ethical fashion represented high costs in terms of the investments in upgrading machinery for the factories and water treatment for the laundering facility. Despite the high costs, the investment of water treatment was partially funded by the EU. The company reaped benefits in terms of reputation as a supplier with a sustainable pedigree which resulted into steady orders and competitiveness. The sewing machines bought for the garment factories do not necessarily produce faster as productivity did not rise considerably. Sewing machinery rather performed a different role this of better quality of manufacture. For example, a new buttonhole fully automatic machine can obviously produce faster, but it can also make better buttonholes which in turn make garments more appealing. The other notion of resource productivity in terms of better production pertains to the fact that since

Continental switched production from fast to slow, the factories needed to produce high complexity garments as upper-middle brands rely heavily on better sewing techniques and craftsmanship in order to justify higher retail prices. This partially concurs with the literature (Martin 2013) where the investments in machinery can enhance the production outcome. However, productivity rises do not necessarily connect to upgraded machinery, especially in a slow production ecosystem. Craftsmanship in stitching requires more time in assembling complex garments. Moreover, smaller orders means that frequent retooling is required, especially in garment factories as higher brands by nature produce smaller quantities which slows down the speed of production.

In view of materials economization, Continental's procurement of yarns for its knitting mills, and procurement of denim fabrics for its woven garment factories were considerably more expensive than in the previous engagement with fast fashion. In this regard, smaller quantities in higher material prices meant that in the cutting room fabric economization was paramount. This is appropriately addressed by the company's operators of computer aided manufacturing (CAM) systems and its cutters. Additionally, nearly all fabric remnants, after order deliveries, were given to garment workers to make their own clothing. Some of them during the time of the authors' visits were making stuffed toys and children's wear not necessarily as a pastime but to clothe their children. This confirms the literature (Organisation for Economic Co-Operation and Development 2011) on the 3R principles as recycling and reusing materials do occur with the exception of reduce in its strict definition. Nevertheless, if the notion of reduce refers to smaller and higher quality production, than the reduce principle applies as well due to smaller quantities of higher value garments.

14 Decent Labor Conditions but no Wage Improvements

In terms of labor conditions, factory workers in all the company's facilities were treated with dignity, which has highlighted the importance social responsibility and also compliance to local labor laws. In all factory floors, there was enough space for every sewing operator to conduct his/her work. Work in progress did not overwhelm the side corridors of production. There was enough ventilation and natural light, and health and safety regulations seemed to apply in all departments. Exits were clearly marked, fire stations were in place, there were no wires hanging from the ceilings, big posters were hanging from the walls to remind all workers of the company's codes of conducts, and there was a resident nurse with a full infirmary room on site. All workers took half hour break for lunch as well as two 10 min breaks for coffee/tea in the morning and late afternoon at the company's cafeteria. Also the factory floor supervisors turned a blind eye for a couple of cigarette breaks within working hours. Work started at seven in the morning and ended at four in the afternoon unless there was overtime of mainly one hour. This rarely exceeded two hours on a day. Every Friday evening, trucks collected the garments for laundering in the facility in Greece. Factories were shut on Saturdays and Sundays complying with the forty hour week stipulated by the EU and local Labor Inspectorates. This does not represent demonstration factories as the literature suggests (Cline 2013). Production was carried out on the specific factory locations without contracting-out. Contract manufacturing to uncertified garment workshops admittedly form part of clothing production in the Balkans but did not occur at Continental as higher quality production could not be easily outsourced because it required specialized machinery and elevated worker skills. The company's garment operations were subjected to strict quality controls on a daily basis by the company's quality controllers and occasionally by external inspectors contracted by buyers. The code of conduct was not imposed by the buying companies as Continental devised its own CSR policies after consultations with lawyers who were experts on European labor law. This partially contradicts the literature (Cline 2013; Siegle 2011) where it is assumed that CSR policies are enforced only by the buyers, and manufacturers have to comply.

Despite the fact that Continental treats its workers in all plants with dignity, the thorny issue of wages rising to living standards did not feature high on the discussions during the interviews with the exception of highly skilled engineers in production, knitting technicians, garment technicians and other skilled personnel. The owners/directors stated that they fully comply with labor laws in all three countries regarding wages, overtime and holidays. However, wages were upgraded to living wages despite higher prices paid for production by the brand owners. Workers enjoyed some non-monetary perks, such as Christmas and Easter food packages, Sunday excursions every two months with all expenses paid for by the company, access to free fabric remnants to make their own clothing after working hours, free refreshments in the factories' cafeterias and first instance private medical cover. This concurs with the literature (Cline 2013; Rankin 2014) as wages in Eastern Europe are not a lot higher in comparison to wages paid in developing countries. This applies especially to women who in Continental's case formed more than 60% of all workers in garment factories.

15 Slowness Prevails in Operations

Slowness principles were demonstrated in Continental's operations since the company switched from fast fashion to a slower mode of production. To be noted, slower production did not mean that workers, supervisors and production engineers took their time in the factories. Production schedules and delivery times were as important as in the previous state. What was different was that speed to assemble and finish garments did not come to the forefront of production decisions as schedules were reconfigured by putting quality of manufacture as the first production priority. This conforms to the natural bio-rhythm promulgated by the slow movement together with other characteristics such as being unhurried, calm and patient (Honore 2004). Through observation during the factory visits, making high complexity garments demanded more careful manufacture which by nature is time consuming.

This unhurried mode of operation was widely acknowledged by the company's production engineers and all workers. Additionally, small production batches required more frequent machinery adjustments as styles may look similar but in reality there were many changes that needed to be carried out for a new style. For example, in terms of the production of a pair of jeans, the thread thickness, the thread colors, sewing machine needles, feet and thread tension, laundering and finishing recipes for the denim fabric could be very different from the previous style.

Continental also sourced yarns and fabrics from European mills knowingly that they were more expensive, but it was acknowledged that quality of materials coupled with inherent durability were very important variables in the manufacturing equation. This was combined with the company's commitment in demonstrating a form of European solidarity for fellow members of the denim supply chain. This slow fashion principle also applied to retaining jobs and skills within Europe in both the company's staff and workers in the extended supply chain (Fletcher 2010). In terms of design, although Continental was not directly involved in design, the company's owners preferred to conduct business with brands conforming to a certain classic approach in their collections. Classic here does not mean "stodgy" in fashion terms, but rather that the brands' designers conceptualized collections which do not easily date. Moreover, in some cases, the company's production engineers and CAM/CAD operators suggested that modifications could improve upon designers' original ideas in the design specification sheets and for economization purposes. Major changes were discussed with the brand owner's design team before proceeding with production. However, smaller changes were carried out without consultations as there was confidence and trust between the firm and retailers because small modifications were for the improvement of the overall product image. For example, small stitching details might be changed or adjustments in pattern grading could be administered for fabric economization purposes or laundering recipes could be slightly tweaked in order to use less water and detergent. This concurs with the slow fashion literature (Fletcher 2008) perhaps not in its stringent interpretation but to the overall notion of the slow movement mindset.

16 Concluding Comments and Perspectives

The detoxification of T&C supply chains from the perspective of reshoring or nearshoring production to Europe can become a reality at least for niche upper-middle brands committed to sustainable manufacturing practices. Nearshoring or reshoring represents a real chance for detoxifying T&C supply chains as the case study demonstrates that from material procurement to laundering and finishing, all production stages can be monitored and implemented successfully by the supply members themselves. The EU's Regional and Coherence funds need to be commended as they play a pivotal role by facilitating sustainable development practices and simultaneously retaining jobs in an admittedly sensitive industrial sector. Furthermore, production networks operating at Europe's fringes demonstrate a renewed dynamism by collaborating closely with brand owners. In this relational scheme, based on socio-cultural solidarity, the parties involved within the supply chain acknowledge that interdependences and trust are paramount factors of success. Despite power asymmetries, the members of the supply chain shall understand their positions as cooperation permeates all operations. Therefore, the net-chain production-distribution-consumption configuration together with the state's influence can be confirmed as all players in the network can influence its outcomes.

The literature and the investigated case study regarding transparent supply chains may form the new imperative. All actors involved from consumers who want to know where, how and who made their clothing, to brand owners and manufacturers whose reputations depending on presenting collections adhering to ethical practices. The ability to trace sustainable supply sources benefits all parties in the net-chain configuration. Additionally, the slow fashion tenets seem to apply in the case study investigated as the company has realized that speed does not represent the predominant priority and instead elected to switch production to a slow mode. High complexity garments manufacture, strict rules on sustainable washing and finishing, and sourcing superior materials and trims represent serious challenges, but this seems to be the way forward for any company wishing to be inserted to upgraded production networks. Alternatives to this model of production mean relegation of European manufacturers to uncertified production methods which currently are not recommended by the EU for Regional funds geared mainly toward sustainable development. Granted that, in the case study under investigation, significant investments in machinery, trade fair participation, staff training are necessary prerequisites, as these actions form the backbone of the company's survival and subsequent competitiveness in the market. However, without these investments, the company would probably have to cease operations as its size could not allow more downsizing.

Last but not least, the thorny issue of workers' living wages as opposed to wages stipulated by Labor Inspectorates seems to remain unsolved as shown in the case study. On balance, upgrading to the elevated status of manufacturing specialized clothing for niche fashion brands can be a viable strategy. Reshoring and sustainable development may represent the two sides of the same coin, and this is also implied by the EU's regional policies. Admittedly, upgrading and transparency combined with slow production can be fraught with difficulties but for manufacturing companies which wish to thrive in this new reality, which may be the only route to competitiveness.

References

Abernathy, F. H., Dunlop, J. T., Hammond, J. H., & Weil, D. (1999) A stitch in Time-learning retailing and the transformation of manufacturing-lessons from the apparel and textile industries. Oxford University Press, Inc., New York

Bair, J., Dickson, M., & Miller, D. (2014) Workers' rights and labor compliance in global supply chains, is a social label the answer? Routledge, Taylor & Francis Group, Oxen

Barrientos, S. (2013) Corporate purchasing practices in global production networks: a socially contested terrain. Geoforum 44: 44–51

Cernat, L., & Pajot, M. (2012) Assembled in Europe—the role of processing trade in EU export performance. The European Commission, Brussels

Cline, E. (2013) Overdressed: the Shockingly High Cost of Cheap Fashion. Porfolio, Penguin Books, New York

Coe, N. M., Dicken, P., & Hess, M. (2008) Global production networks: realizing the potential. Journal of Economic Geography 8: 271–295

Creswell, J. W. (2009) Research design: qualitative, quantitative and mixed methods approaches (3rd ed.). Sage Publications Inc., California

Curwen, L., Park, J., & Sarkar, A. (2013) Challenges and solutions of sustainable apparel production development. Clothing & Textiles Research Journal 31(1):32–47

De Brito, M., Carbone, V., & Blanquart, M. (2008) Towards a sustainable fashion retail supply chain in Europe: organization and performance. International Journal of Production Economics 114: 534–553

Dicken, P. (2003) Global Shift (4th ed.). Guilford Press, New York

Donner, S. (2015) Why we need the next-to-impossible 1.5 °C temperature target. https://www.theguardian.com/environment/climate-consensus-97-per-cent/2015/dec/30/why-we-need-the-next-to-impossible-15c-temperature-target

Dyer, J. H., & Singh, H. (1998) The relational view: cooperative strategy and sources of inter-organizational competitive advantage. Academy of Management Review 23(4): 660–679

Ellram, L., Tate, W., & Peterson, K. (2013) Offshoring and reshoring: An update on the manufacturing location decision. Journal of Supply Chain Management 49(2): 14–22

European Commission (2015) Directorate-General for trade. http://ec.europa.eu/regional_policy/en/funding/cohesion-fund/

European Parliamentary Research Service (2014) Reshoring of EU manufacturing. Strasbourg European Parliamentary Research Service http://www.europarl.europa.eu/EPRS/140791REV1-Reshoring-of-EU-manufacturing-FINAL.pdf

Fernandes, V. (2012) Understanding reverse offshoring: a theoretical and empirical study. http://www.esc-larochelle.fr/eng/Research-Faculty/Faculty/Faculty-and-Research-departments/Faculty-and-Research-departments/Strategy-Department/FERNANDESFERNANDES,%20V.%20(2012).%20Understanding%20Reverse%20Offshoring:%20A%20Theoretical%20and%20Empirical%20Study.%20The%20European%20Business%20Review-Valerie

Fletcher, K. (2008) Sustainable fashion & textiles: design journeys. Earthscan, London

Fletcher, K. (2010) Slow fashion: an invitation for systems change. The Journal of Design, Creative Process & the Fashion Industry 2(2): 259–265

Friedman, V. (2010, February 5th). Sustainable fashion: what does green mean? Financial Times.

Gereffi, G., Humphrey, J., & Sturgeon, T. (2005) The governance of global value chains. Review of International Political Economy 12(1): 78–104

Gereffi, G., & Lee, J. (2012) Why the world suddenly cares about global supply chains. Journal of Supply Chain Management 48(3): 24–32

Germani, M., Mandolini, M., Marilungo, E., & Papetti, A. (2015) A system to increase the sustainability and traceability of supply chains. Procedia CIRP, 29: 227–232

Goworek, H. (2011) Social and environmental sustainability in the clothing industry: a case study of a fair trade retailer. Social Responsibility Journal 7(1): 74–86

Greenpeace International. (2012) Toxic threads: the big fashion stitch up. Greenpeace International, Amsterdam

Groom, B., & Powley, T. (2014, March 3) Reshoring driven by quality, not costs, say UK manufacturers. Financial Times

Hargreaves, B. (2014) Made in Britain. Confederation of British Industries (CBI), London

Hargreaves, S. (2013) Manufacturing jobs dry up. CNN, Atlanta

Honore, C. (2004) In praise of slowness, challenging the cult of speed. Happer Collins Publishers, New York

Lavie, D. (2006) The competitive advantage of interconnected firms: a extension of the resource-based view. Academy of Management Review 31(3): 638–658

Lubin, D. A., Esty, D., & May, C. (2010) The sustainable apparel imperative. Harvard Business Review, 88: 42–50

Luginbühl, C., & Musiolek, B. (2014) Stitched up: poverty wages for garment workers in Eastern Europe and Turkey. Clean Clothes Campaign http://digitalcommons.ilr.cornell.edu/cgi/viewcontent.cgi?article=2842&context=globaldocs

Martin, M. (2013) Creating sustainable apparel value chains: a primer on industry transformation. Impact Economy, Geneva

Mellery-Pratt, R. (2014) A row of opportunity, Part 1, Business of Fashion https://www.businessoffashion.com/articles/intelligence/row-opportunity-part-1

Mellery-Pratt, R. (2015, Jan. 12) Made in Portugal is on the rise. https://www.businessoffashion.com/community/voices/discussions/does-made-inmatter/made-portugal-rise

Oppenheim, A. (1992) Questionnaire design, interviewing and attitude measurement (1st ed.). Cassell Wellington House, New York

Organisation for Economic Co-Operation and Development (2011) Resource productivity in the G8 and the OECD. Organization for Economic Co-Operation and Development, Paris

Pickles, J., & Smith, A. (2011) Delocalization and persistence in the European clothing industry: the reconfiguration of trade and production networks. Regional Studies, 45(2): 167–185

Pickles, J., Smith, A., Buck, M., Roukova, P., & Begg, R. (2006) Upgrading, changing competitive pressures, and diverse practices in the East and Central European apparel industry. Environment and Planning A, 38(12): 2305–2324

Rankin, J. (2014) Fashion brands violate labour laws in Eastern Europe, NGO report finds. The Guardian https://www.theguardian.com/business/2014/jun/10/fashion-brands-violate-labour-laws-eastern-europe

Reimer, S. (2009) Geographies of production II: fashion, creative and fragmented labour. Progress in Human Geography, 33(1): 65–73

Remy, N., Speelman, E., & Swartz, S. (2016) Style that's sustainable: a new fast-fashion formula. McKinsey & Company, Sustainability & Resource Productivity http://www.mckinsey.com/business-functions/sustainability-and-resource-productivity/our-insights/style-thats-sustainable-a-new-fast-fashion-formula

Seuring, S., Sarkis, J., Muller, M., & Rao, P. (2008) Sustainability and supply chain management Journal of Cleaner Production, 16(5): 1545–1710

Shih, W. & Agrafiotis, K. (2015) Competitive strategies of new product development in textile and clothing manufacturing. The Journal of the Textile Institute, 106(10): 1027–1037

Siegle, L. (2011, 8th May) Why Fast Fashion is Slow Death for the Planet. The Observer, p. 35

Siegle, L. (2016, March 13th) Is it time to give up leather? The Guardian https://www.theguardian.com/fashion/2016/mar/13/is-it-time-to-give-up-leather-animal-welfare-ethical-lucy-siegle

Sirkin, H. L., Zinser, M., & Hohner, D. (2011) Made in American, again: why manufacturing will return to the US. The Boston Consulting Group https://www.bcg.com/documents/file84471.pdf

Smitha, E. (2011) Screwing mother earth for profit. Watkins Publishing, London

The Economist (2013) Reshoring manufacturing coming home. The Economist http://www.economist.com/news/special-report/21569570-growing-number-american-companies-are-moving-their-manufacturing-back-united

Tokatli, N. (2008) Global sourcing: insights from the global clothing industry-the case of Zara, a fast fashion retailer. Journal of Economic Geography, 8: 21–38

United Nations Global Compact Office (2014) A guide to traceability: a practical approach to advance sustainability in global supply chains. United Nations, New York

Van der Pols, D. (2014). CBI special topic: supply chain trends in the apparel sector. Ministry of Foreign Affairs, the Netherlands, Amsterdam

Wingard, C., & Connerty, M. (2014, June 4th) The rebirth of US manufacturing: myth or reality? https://hbr.org/2014/06/the-rebirth-of-u-s-manufacturing-myth-or-reality

Worland, J. (2015, Dec. 13th) What to know about the historic 'Paris Agreement' on climate change. http://time.com/4146764/paris-agreement-climate-cop-21/

Yin, R. (2009) Case study research: design and methods (4th ed.). Sage, California

Za, V. (2014) Euro zone companies come home as Asian costs rise. http://www.reuters.com/article/us-eurozone-reshoring-idUSKCN0JF21Q20141201

Detoxifying Luxury and Fashion Industry: Case of Market Driving Brands

Ivan Coste-Manière, Hamdi Guezguez, Mukta Ramchandani, Marie Reault and Julia van Holt

Abstract The current chapter encompasses the distinction between market driving and market driven brands in the sustainable luxury and fashion industry. In particular, how brands are detoxifying their manufacturing, supply chain distribution and market along with the methods they have adapted to differentiate themselves in this process. In the long run, how these methods economically impact the brands and consequences followed from the consumer's perspectives are explored. Stella McCartney published for the first time its environmental activity statement for the last three years of the brand in 2016. The fashion designer is also famous for her commitment to sustainability and her environmental friendly creations. Sustainability and detox are at the heart of debates within modern society and

---

In Memoriam our friend, student and much more, our beloved Cédric Laguerre.

I. Coste-Manière (✉) · H. Guezguez · M. Reault
SKEMA Business School, Suzhou, China
e-mail: ivan.coste@skema.edu

I. Coste-Manière
Sil Innov & Eytelia, Courcelles, Belgium

I. Coste-Manière
SKEMA Business School, Sophia Antipolis, France

H. Guezguez
Danah-Industry, Shanghaï, China

M. Ramchandani
Marketing Consultant, Olten, Switzerland

M. Ramchandani
United International Business School, Zurich, Switzerland

M. Ramchandani
Neoma Business School, Reims, France

J. van Holt
International School of Management, Cologne, Germany

J. van Holt
Eytelia, Courcelles, Belgium

© Springer Nature Singapore Pte Ltd. 2018
S.S. Muthu (ed.), *Detox Fashion*, Textile Science and Clothing Technology,
DOI 10.1007/978-981-10-4783-1_2

represent the main stake for the future, the organization of COP 21 in 2015 in Paris being one of the latest examples. However, when we talk about luxury industry, especially fashion, sustainability appears to be difficult to associate to this sector. Indeed, luxury is representing a kind of "unfair" and fast moving consumption, with some ostentation, while ecology involves the protection of resources and durable models. But sustainability and detox are representing a tremendous opportunity for luxury fashion brands. Luxury and fashion brands' strategies need currently, within an increasingly competitive market, to shift from such paradoxes and to capitalize on being the greatest opportunity to dodge "green washing" advocacy communication.

Keywords Luxury · Fashion · Brands · Sustainability · Detox · Resources · Development · Strategy · Stella McCartney

1 Introduction

The shift towards a sustainable mindset in daily life is omnipresent these days. But in the fashion business and especially in the luxury industry, exists still a huge lack of transparency concerning this topic. Many companies have production sites located in Asian countries, such as China or India. Especially in such locations, the value chain is often not traceable anymore, and brands can tell their customers whatever they want to claim they are sustainable. This process of greenwashing is more and more feared by the consumers.

The following sections are dealing with this subject of 'real' versus 'wannabe' sustainability, and is therefore firstly analyzing two examples of market drivers. The very different approaches of Stella McCartney on the one hand and Valentino on the other hand will be compared, to give an example for the luxury industry.

But what makes Stella McCartney to be accepted as sustainable in the consumers' minds? This will be explained in the following sections, by giving a closer look to luxury customers' expectations and by analyzing who the drivers of the change might be. What are the different cultural approaches to that topic in China and in Western Europe? Do luxury consumers all think in the same way or are adaptions necessary?

Additionally, the term of 'detoxification' will be further explained. With Greenpeace, as a major NGO and leader of several anti-toxin campaigns, this part of the chapter will explain why detox is needed in the fashion and luxury industry and how the consumers are involved.

In the last part of the chapter it will be shown how to avoid greenwashing. It will sum up why a good corporate social responsibility will soon be crucial for luxury and fashion brands, and that there better be some truths behind those tactics.

2 Methodology

In this chapter the impact and concept of detoxifying the fashion and luxury industry is explained. Therefore, in the first step secondary literature has been consulted, such as articles, journals, blogs and company websites. The topic of sustainability furthermore goes hand in hand with the use of environmental reports. Here we included material from the brands as well as NGO reports drawn into consideration.

Additionally, as a primary source, a questionnaire and an interview was conducted. In this way, the most-up to date consumer-driven data was collected, to give an idea of the current scenario and the future directions.

Finally, all the primary and secondary collected data was drawn into consideration for our analysis and results. In summary, the end of the chapter develops a set of recommended actions to help brands implement a better CSR, without greenwashing their portfolio. Or are luxury and sustainability really an antithesis with each other?

3 Market Driving Brands

Only recently, luxury brands started to integrate sustainability in their value chain. There are two kinds of brands in the field of sustainability existing: Firstly, the ones who drive the markets, which are in the same time brands that totally adapted a sustainable identity to their core values.[1] Who are those brands and why are they considered to be market driving? This innovative way of thinking influences the whole market. Hence secondly market driven companies are those, who follow the rules of the preferences in this given structure (Jaworski et al. 2000).

Stella McCartney is one of the most famous market driving sustainable luxury brands. Her main approach to fashion is to provide excellent tailoring and to design sexy but confident pieces for women. In the same time the company tries to be environmentally and ethically correct. In these values the aspect of responsible sourcing and supplying is very important for the brands beliefs (Kering 2016). Instead of using leather or fur, the vegetarian designer provides high quality alternatives to her customers, such as a new material called 'Eco Alter Nappa' (Stella McCartney 2016).

With this foregoing description, the sustainability is also a prospect for innovation. 'Eco Alter Nappa' is made from polyester and polyurethane which is coated with 50% vegetable oil. This mix of materials form an alternative for leather and uses less mineral oil than normal coating. Stella McCartney uses this material for example for her iconic Falabella Bags (Stella McCartney 2016). This model reached a well-known and well-accepted status in the luxury fashion world.

---

[1]http://www.huffingtonpost.com/misha-pinkhasov/from-sustainable-luxury-t_b_7069074.html.

But originally the innovations of sustainability need a group of early adapters to be willing to get started with this 'new' form of luxury.

Thus, it requires the right psychological convincing of the consumer. What made the Stella McCartney Customer step away from the traditional luxurious leather materials? What makes the brand communication so authentic is also the fact that it admits flaws:

> We know that we are not perfect. We also know that sustainability isn't just one thing. (…) In many ways we are just beginning our journey towards becoming more sustainable, but we are dedicated to continuing our work towards being able to replace what we have taken from the environment. (…) We will probably never be perfect, but you can rest assured that we are always trying. (Stella McCartney 2016).

With this statement, Stella McCartney tries to modify the customers' expectations. Till then, the mindset always includes an 'all-or-nothing'-approach, in which the customer would not allow the brand to step slowly into sustainability. But this fact only leads back that companies were often only 'greenwashing' their images. In a sustainability approach in which the customer sooner or later learns that literally something hazardous was hidden behind every brand, it is a logical consequence that today's customer has a hard time to really trust a company who only state that it is sustainable. At this point, Stella McCartney's approach catches the trend: The designer never claims to be 100% sustainable. With admitting that, she makes herself more trustworthy for an authentic green-customer.

It is considered as a state of the art by now, that luxury brands discovered the potential of Corporate Social Responsibility for their brand identity. But Stella McCartney's approach seems to be more honest one. The whole corporate identity and communication strategy of the brand is built around the image of sustainability and vegetarianism, and thus it is used as a differentiator of the brand. And it seems to be working out: 2015 has been the brand's best year so far. In the last three years, it's ecological footprint decreased by 35% per kg of used material (Van Looveren 2016).

On the other hand, the luxury house **Valentino** is in the meantime also considered to be detox-driving. Yet, the brands approach to it, seems to be different than Stella McCartney's. What strikes the eye in Valentino's sustainability approach is firstly: nothing. The consumer cannot find a hint to the topic on a first glance on the brands homepage. Only when scrolling to the bottom of the page and clicking on the link 'corporate information' one can find goals of the detox-campaign (Valentino 2016).

In the 'Valentino Fashion Group Detox Solution Commitment' the brand sets the goal to get rid of all hazardous chemicals in its whole value chain until 2020. The document is set up as a formal contract and can be downloaded by every interested customer from the brands website. As such, the commitment refers to the consumers 'right to know' about the chemical substances the brand uses. It furthermore promises to eliminate those substances one after another and to be transparent enough to inform the public about this process (Valentino 2016).

And yet, this Detox Commitment also indicates nothing more but the will to change. In contrast to Stella McCartney, Valentino never refers to itself as a 'sustainable' or 'ethic' brand. But through making their will to change formally public, the brand shows their interested customers that it is on a good path. Both brands clearly state that the detox vision is a long-term commitment, which is not easy to achieve. Stella McCartney tries to achieve a totally detoxed supply chain very obviously, while Valentino does it rather quietly, without making their actions obvious.

3.1 Communication Sense

Brands are always connected to an evoked set of perceptions of their customers. Especially in the luxury industry the customer's preferences are highly linked to emotions (Beiz 2006). And yet, so is sustainability. What would it do to the brand identity of Valentino if it would communicate more its' will to move into a sustainable direction? The brand Valentino is at this point an excellent example for many luxury brands. Valentino, especially, is overall linked to high class femininity, seduction and the brands legendary haute-couture evening robes. Nothing in the brands image could be generally connected to the image which sustainability has since now: A more honest, more down to earth, more unromantic image. It seems therefore legitimate that the dreamy luxury brand like Valentino does not clearly communicate their sustainable commitments.

In the last few years, the number of market driving luxury brands in the field of sustainability barely rose recognizable. Other brands are not yet this far. For them sustainability is a differentiated part of their business: A part which is nice to have, to generate sales and to create a good image. But those brands treat their corporate Sustainability as a part of their marketing and communication process (Pinkhasov 2015). This often happens because of public pressure which is nowadays put on international businesses. An ecological or social scandal would have a huge negative impact on the reputation of the brand and the buying behavior of the consumer (Merck and Fleischer 2013).

4 Survey

We conducted a survey to show the customers reaction to the increasing impact of sustainability in the luxury and fashion industry. The questions were sent to a non-representative group of consumers in China and Western Europe (n = 282), to compare both cultural approaches to this delicate topic. In total 118 Chinese participated. On the other hand, 137 people located in western Europe participated. Herewith it is important to mention that the majority, 94% of the participants come

Table 1 Demographic profile of the sample

	Variable	Total		China		Western Europe	
		n	%	n	%	n	%
Gender	Male	101	35	45	37	50	37
	Female	181	65	73	63	87	63
Occupation	(Self-) employed	128	45	62	53	53	39
	Student	140	50	53	45	75	55
	Not working/retired	14	5	3	2	9	6

from Germany and France. Both these countries have a well-developed and growing market for sustainable luxury and fashion goods. The total sample demographical profile is shown in the Table 1.

The survey included questions regarding participant`s opinion about their general interests in buying sustainable luxury and fashion products; if it makes a difference for them regarding how the luxury and fashion products are sourced sustainably or not; and if they would consider buying more sustainable luxury and fashion products if the brands would provide them. Additionally, we measured their feelings (*positive/neutral/negative/inconvenient*) when buying a sustainable luxury and fashion product. Moreover, they also answered in their opinion who do they consider responsible for sustainable products (*Consumers/Government of the production country/Government of the selling country/The brands/The suppliers/Others*). Finally, they answered how can luxury brands improve their reputation in sustainability (*Publish an environmental activity statement/Being tested by NGOs on a regular basis/Making their supply chain transparent/None of these, luxury and sustainability together will never be trustworthy/I don't know/Other*).

A study of KPMG in 2012 showed that sustainability attempts are often driven from a governmental side and by laws. Our survey could not confirm that: In total, most of the consumers see the responsibility rather on customer-, brand- and supplier-side than as a governmental issue.

Often 'being sustainable' is not what it seems like. For example, are synthetic fabrics not necessarily more damaging for the environment than natural products. If we just think of the water usage of a pair of jeans compared to a textile made of old plastic bottles, this point gets clear. But the PR-Department of most companies can simply make more nice images out of healthy glowing cotton fields than of non-natural sources. The true picture of sustainability is often disturbed by what media channels make of it and by what the public wants to believe (Pinkhasov 2015).

This emphasizes the role of the customer. Sustainable market driving brands are more likely to be bought by a group of people who *know* and who *care*. The next section will analyze how consumers are affected by sustainable driven luxury brands and if it is going to change their perspective on the industry.

4.1   The Customer's Perspective?

Why do some brands practice market driven greenwashing? Because the image of sustainability by itself still has a problem. From the business side, sustainable methods are generally reckoned to be more complicated, slower and for the most important part more expensive (Pinkhasov 2015). But also, consumers tend to think that way. Sustainable fashion and luxury has moreover the image of being high priced due to more complicated production. Hence 'buying sustainable' is seen as rather inconvenient even though linked with the acknowledgement of the positive idea behind it (Petersen 2015).

In fashion, rather than in luxury, the customers are getting more sensitive. In response to the recent catastrophes, such as building collapses in Bangladesh, consumers are starting to pay attention (Merck and Fleischer 2013). Our survey resembles this trend: Most of the test group is interested in sustainable produced fashion and luxury goods (70%), as well as they are interested in buying more of those products if luxury brands would provide them (68%).

Yet, for most of them it does also make a difference if this product would be sourced sustainable. That means that either in a positive or negative way their buying decision would be affected by knowing that a piece is sustainable. Thus, we also asked for the consequences. The results are striking: Only 7% of the participants linked something negative and inconvenience with buying sustainable, and 57% only bought sustainable to feel better.

The benefit for the customer must be more than only a clean consciousness and the fulfilment of moral and ethical obligations (Gutjahr 2013). Brands still need to provide the decision driving benefits, such as function and design, which are still more important for the customers (Beiz 2006). The environmental commitment of the brand is not a decisive factor, according to a study of Achabou and Dekhili (2012). This study furthermore shows, that consumers are by far more likely to buy sustainable luxury products, if those goods have an equivalent high intrinsic quality standard to former bought, non-sustainable products (Achabou and Dekhili 2012).

The goal is hence, to sell sustainable luxury and fashion with no difference to conventional, former goods. At least, 21% of our test group claimed that it would make no difference to them, but the tendency for a rise of this number is not unlikely. If more brands like Valentino sell their sustainable luxury items without further communication, it will not much more occur to the customer's consciousness. This verifies the percentage of people who would keep buying 'their' luxury brands if it would start selling sustainable goods: 62% would be happy about it, 10% would feel positive till neutral about it and for 25% it would not even make a difference. Only barely 3% of the asked demanded the brands to change nothing.

But who should lead the action for the change? As mentioned most participants in our survey answered that the brands and the consumer should take the initiative. But with respect to the cultural background, the answers differed. For the Westerners, these responsibilities were very strongly defined: A significant high amount of them see the responsibility on the brands' side, followed by the consumer themselves. These two were followed by the supplier-industry with still nearly half of the people's votes (46%). This finding proves what is also widely spread in literature and in the web: Without these three parties' cooperation, the shift to sustainable luxury will 'simply remain an ideology' (Guldager 2016).

But on the Chinese side, this survey question showed a major difference: Firstly, the brands and the consumer were too, the prior responsible, but with less significance than for the western. And secondly it is maybe not surprising to see, that over half of the Chinese participants saw also the government of the production country on duty. The following section will give some explanations about China as a production country and why detox is important.

5   The Term of Detoxification

Detoxification means getting rid of toxins and other hazardous chemicals in consumer's products. In 2011, a detox campaign was launched by Greenpeace to raise awareness of the toxic products and exposed the links between global clothing brands, suppliers, and toxic water pollution around the world. The logo contained a Chinese sign "Shui" meaning water as the alphabet X in DetoX. According to Qian (2016), the use of the Chinese sign was an indicator of the main hazardous chemical problems associated with China. As it is the biggest manufacturer in the world with over 25 thousand chemical companies.

Especially the textile industry in China is responsible for toxic water pollution. The suppliers located there deal with a wide range of hazardous chemicals for clothing, manufacturing and processing. This leads to a high amount of industrial discharges, which contaminates the Chinese environment as well as the human health (Greenpeace 2012). The past has shown that most brands do not even feel responsible for what is happening in manufacturing companies in their supply chain.

A campaign was set up, from the non-governmental organization to put pressure on some of the biggest fashion companies. Greenpeace demands an elimination of all hazardous products from companies like Nike, H&M, Valentino and Burberry. Those and 14 other companies already signed a public commitment to create concrete steps towards more transparency and less hazardous substances (Greenpeace 2016).

Greenpeace believes that such big global players should lead the whole industry to a cleaner production. The campaign focuses very much on the point of production, but less on the issue of the supply chain. The NGO even asks the brands to redesign their products if it is not possible to get rid of hazardous ingredients otherwise.

Yet, China as a production country is very much on the map of the detox-movement. Due to a lack of governmental restrictions a lot of hazardous accidents are happening in the world's biggest chemical industry. Greenpeace found out, that the environment of the country has to deal with an average of 29 chemical accidents per month, of whose many lead to bad consequences such as injured workers, death and huge ecological problems (Qian 2016).

By January 2020 the NGO wants big brands to completely "phase out the use of toxic chemicals from their global supply chain and products" (Greenpeace 2016). At the same time it is very difficult to get information about those supply chains. As for China, for example, there are still a lot of white spots on the map. And the Chinese government has shown not much interest to change that status (Qian 2016). We found similar results reflected in our survey: That is why Chinese see the governmental responsibility much as much higher as the Westerns do.

Greenpeace, furthermore, sees the responsibility for action not only at the companies, but also on the governments site. They ask for a set of 'chemicals management policies' in the country of production. It is seen to be as questionable that a set of new rules in one or two supplying countries could make a noticeable change in the industry.

Last but not least Greenpeace emphasizes the social media hashtag #PeoplePower. But the NGO also underlines, that the consumer can only make a change when a larger group of consumers is shifting their mindset and raise their voices (Greenpeace 2016). The consumer is asked to take action in form of putting pressure on brands and governments, but it is not described how a normal citizen shall influence such big industries.

The Detox campaign as a sustainability movement has a comparable high chance to touch the end-consumer directly. The trend started as a personal food-diet in which participants shall get rid of toxins in their body. Detox is about the consumer's direct lifestyle. Various studies have shown, that the impact of sustainable offers is especially high when it comes to the consumer`s instantaneous lives. Health issues are building an important line of argumentation in the customer's decision journey. Benefits are considered as weaker, when they do not attach the direct lives of the consumer, such as the climate change or the deforestation. Those tragedies are well aware, but less touchable for the consumer (Gutjahr 2013).

6 Interview

We conducted an Interview with Arianna Bolzoni and Fahid Jaaouan founder of OPS-ARTIFEX to understand further about the current issue within the field of sustainable luxury. Following are the questions and responses from the interview.

1. **Can you tell us a little about your backgrounds and introduce us a bit your company**.
   We are called OPS-ARTIFEX (in Latin it means the support of the skilled) We are a fashion design studio specialized in denim.

2. **What, in your opinion, is the most important thing to look for in ethical fashion?**
   It has to be pretty, It has to look cool, it has to be in trend, it has to be commercial. The ethical means to me sustainable product, where water consumption is reduced, chemicals dyes are not hazardous for our planet.

3. **What do you think is the biggest obstacle to becoming a more sustainable and less harmful industry?**
   The price! Brands are too greedy! They rather give incentives to buyers and sourcing managers who help them make bigger margins. I believe government should step up and strategies such as implementing high taxes to brands who still uses harmful product. We all know which chemicals are harmful and hazardous, hence we should just get high tax on all those chemicals and reward the brands who are sustainable by deducting taxes. Just like when using solar panels for your house, you are allowed to deduct the related charges from your taxes. That's a win-win strategy.

4. **How do you think we can increase the use of sustainable materials in the fashion industry?**
   As mentioned before, the only motive for company is profit, therefore by Increasing taxes of all the products who aren't sustainable, you will get a positive impact very quickly.

5. **How do you choose what brands to work with?**
   We work with all the brands without any exception. We usually find them solutions to create beautiful collections, matching the trends and their market in a sustainable way, keeping the same budget.

6. **How have you seen the sustainable fashion industry change/develop in the past five years?**
   Yes indeed! it is changing a lot! Many global campaigns focus on sustainability. Groups like Greenpeace has challenged many fast fashion and mass market brands to use zero hazardous product by 2020.

7. **Have you seen more interest from consumers lately?**
   Frankly no, at least not enough to make any changes. Consumers are driven by prices and trends!
   I believe we will be able to see change in consumer's purchase behaviour, if the big players would change their manufacturing process to an eco and social friendly one.

8. **In your opinion, will it be the consumer who will facilitate the change in the supply chain, or will it be the design and manufacturing industry?**
   Being in the industry for more than a decade, the only engine to change is R&D, design and manufacture.

9. **People charge that ethical and sustainable fashion is too expensive. How do you respond to that?**
   It's true! Once again! Sustainable means, environment, social and economy. If those 3 pillars aren't respected then it is not sustainable.

10. **Fostering sustainability and responsible consumerism is on top of the agenda for many players in the denim industry. What other brands do you most respect for driving sustainable initiatives.**
    Nudie jeans are doing great sustainable products, but by far my support goes to Candiani Denim! An Italian denim mill known to be the greenest denim mill in the blue world.

The interview provided us with deeper understanding of the subject. The next section describes further about the CSR aspect in sustainable luxury brands.

7 Opportunity and Potential of 'Good' CSR

The sustainable luxury image has to shift itself to a more self-evident luxury and fashion industry. A detoxed luxury and fashion industry must not be part of a quite consciousness but simply a part of the aesthetics and dreams that brands are selling. In the last few years, literature was mostly claiming that sustainability has to be deeply rooted in the brands identity (Gutjahr 2013). But market driving brands like Valentino are the best examples that this assumption is no longer the only truth.

The perceived cognitive value of a brand is the connected with the actual bundle of advantages of a brand. Therefore, this cognitive knowledge is most likely subjective and linked with the customer's emotions (Meffert et al. 2012). Those emotions, however, most not necessarily be linked to 'doing good' or 'buying eco-friendly'. Instead of using guilt as an argumentation line for sustainable fashion and luxury industry, the brand communication should rather implement this strategy as a natural part of their value chain (Petersen 2015).

In the brand communication, authenticity is still the most important aspect. Only when the customers recognize an honest approach to a detoxed value chain, the brands can avoid the accusation of greenwashing (Gutjahr 2013). We asked the consumers of China and Western Europe how luxury brands could improve their reputation in sustainability. With some distance 68% of the consumers would feel better if luxury brands would make their supply chain transparent on their website, for example. Especially for the western world, this was an important point. When the country of production is physically as well as culturally far away, it is hard for this demographic group to calculate the actual Status Quo.

This policy was followed by the proposed measures of publishing an environmental report and being tested by NGOs on a regular basis. NGOs are well trusted by the consumers. But concerning their sustainability-communication, NGOs play an important role for the brands, too. In the reputation-process they plan a significant role as an intermediary between brands and consumers (Fabisch 2010). In the past NGOs like Greenpeace were rather feared by brands, because they reported about illegitimate VERHÄLTNISSE in their supplier's production sides.

To take apart real detox-attempts and greenwashing, it is in the brands responsibility to educate and empower their customers. This might result into changing existing buying habits (Achabou and Dekhili 2012).

The problem of greenwashing occurred parallel with the trend to publish a Sustainability Performance Report (KPMG 2012). When publishing those reports without further proof, the companies would obviously embellish their detox attempts. 52% of our participants claimed that a published report would improve the image of a brands sustainability approach. If that shall work out, the brands have to provide transparent and comprehensible data for their customers. It is a quite likely scenario that only companies and brands that will provide an authentic sustainability approach will gain the trust of their customers on the long term (KPMG 2012).

Brands should see the detoxification as a part of innovative branding (Gutjahr 2013). Stella McCartney is a pioneer in this field. Through her sustainable marketing communication, she gives the customer the feeling to be able to solve ecological and ethical problems through buying her innovations (Gutjahr 2013). Herewith the customer experiences a double self-improvement: His reward center requites him for buying a design which a higher social class would consider as beautiful and at the same time bought (exaggeratedly expressed) the feeling that he or she has saved the world.

And yet luxury goods and sustainable production share some common aspects: They require extraordinary materials and production processes, and both have the reputation to be rather rare since now (Petersen 2015). Within our test group, only a significant small number of people would claim that luxury and sustainability won't go well together (6% in total).

Various former studies have shown that luxury if often possessed because consumers accredit it with a higher quality than low-priced products (e.g., KPMG 2009). Within the test group who are already loyal luxury customers, 90% were convinced of the better quality. This perception is the grip-point for sustainability. Because how high can a quality of a product be, as long as there are hazardous chemicals still in the materials?

Luxury brands more than fashion brands have the power to manifest sustainable buying as a lifestyle choice (Petersen 2015). Even though luxury and fashion seemed to be a paradox for a long time, now the question is no longer *if* but *how* to implement sustainability in the brand management.

Luxury and fashion brands need to take action for consumers.In order to maximize the perceived value in the context of sustainability excellence, the brands need to figure out what level of sustainability awareness their customers want to see. Adjusted to this the supply-chain needs to be examined and taken into operational strategies. Only in this way brands can create deeper value and hence clearly differentiate "the 'real green' from the green-washing" (Sauers 2010).

8   Conclusion

What strikes the eye, is that there is still a very small amount of true market driving brands in the field of sustainable luxury and fashion. This is not necessarily a confession of failure for the industry, but rather a statement of how difficult it is for brands to jump into sustainability. There is no such thing as the perfectly detoxed brand in the market yet, but under governmental and public pressure, more and more of them are trying the shift.

For this tactical need, the brands identity has to evolve around the idea to be as good to nature and humankind as possible. Since fast fashion is all about fast-lane consumerism, it seems to be more in the nature of luxury to develop a true approach to sustainability. A turn to 'slow fashion' would enhance the aspect of luxury, and develop the aspect of sustainability (Coste-Manière et al. 2015). This argument was furthermore backed up by the survey made, which showed that most participants are not only interested in buying sustainable, but would moreover even like to see 'their' luxury brands shifting in this direction.

Additionally, the survey was able to show the differences between the cultural mindsets about sustainability. It got clear, that compared to Western Europeans, Chinese see responsibilities and measures on a slightly different focus. Therefore, it is important for brands to educate international customers with adapted cultural styles.

What concluded clearly from the analyzed materials of the survey and the interview, is that brands have to set a good example. But to avoid greenwashing, it is inevitable for brands to work with customer education as well as with restrictions for their suppliers and even the governments. Without those cooperations, the shift to sustainability will be hardly made within the industry.

Appendix: Survey Answers

Questions	Answers	Total		China		Western Europe	
		n	%	n	%	n	%
Are you interested in buying sustainable luxury and fashion goods?	Yes	197	70	71	60	110	80
	No	56	20	34	29	13	10
	Maybe	29	10	13	11	14	10
Does it make a difference for you if a luxury or fashion product is sourced sustainable or not?	Yes	170	60	68	58	86	63
	No	82	29	38	32	36	26
	Maybe	30	11	12	10	15	11
Are you interested in having and/or buying more sustainable luxury products if the brands would provide them?	Yes	192	68	67	57	109	80
	No	59	21	37	31	14	10
	Maybe	31	11	14	12	14	10

Questions	Answers	Total		China		Western Europe	
		n	%	n	%	n	%
In your opinion: What are the consequences for you when buying a sustainable luxury product?	A good feeling (positive)	161	57	70	60	75	55
	Positive—neutral	39	14	8	7	29	21
	No difference (neutral)	60	21	28	23	27	20
	Neutral—negative	2	1	1	1	1	1
	Inconvenience	20	7	11	9	5	4
Does it make a difference for you if a luxury or fashion product is sourced sustainable or not?	Yes, I would like it (positive)	173	62	79	68	81	59
	Positive–neutral	28	10	3	3	22	16
	No, it would not make a difference (neutral)	70	25	31	26	32	23
	Neutral-negative	1	0,5	1	1	0	0
	Yes, I don't want them to change anything (negative)	8	3	4	3	2	2
In your opinion: Who has the responsibility for sustainable products?	The consumer	161	57	63	53	87	64
	The government of the production country	123	44	61	52	49	36
	The government of the selling country	94	34	42	36	40	29
	The brands	217	77	82	70	115	84
	The suppliers	138	49	61	52	63	46
	Others	3	1	2	2	1	1
How do you think luxury brands could improve their reputation in sustainability?	Publish an environmental activity statement	146	52	61	52	67	49
	Being tested by NGOs on a regular basis	141	50	63	53	68	50
	Making their supply chain transparent	192	68	72	61	104	76
	None of these, luxury and sustainability together will never be trustworthy	17	6	14	12	2	2
	I don't know	25	9	13	11	11	8
	Other	9	3	4	3	5	4

References

Achabou M A, Dekhili S (2012) Luxury and sustainable development: Is there a match? In: Journal of business research 66: 1896–1903

Beiz, C. (2006). Spannung Marke. Springer, Wiesbaden

Coste-Manière I, Ramchandani M, Chhabra S, Cakmak B (2015) Long-Term Sustainable Sustainability in Luxury. Where Else? in: Gardetti M-A, Muthu S (ed) Handbook of Sustainable Luxury Textiles and Fashion, 2nd edn. Springer, Singapore, p 17–35

Fabisch N. (2010) Nachhaltigkeitsmarketing als innovativer Strategieansatz. In: Loock H, Steppeler H (ed)Marktorientierte Problemlösungen im Innovationsmarketing. Springer, Wiesbaden, p 461–478

Greenpeace (2016) Detox Campaign. http://www.greenpeace.org/international/en/campaigns/detox/fashion/about/. Accessed 24 Nov 2016

Greenpeace (ed) (2012) Dirty Laundry. http://www.greenpeace.org/international/Global/international/publications/toxics/Water%202011/dirty-laundry-report.pdf. Accessed 24 Nov 2016

Gutjahr G. (2013) Markenpsychologie. 2nd edn. Springer, Wiesbaden

Jaworski, B, Kohli, A.K Sahay A. J. (2000) of the Acad. Mark. Sci. (2000) 28: 45

Kering (2016) Stella McCartney LINK, Accessed 25 Nov 2016

KPMG AG (ed) (2012) Trends im Handel 2020. Köln/Hamburg

KPMG AG (ed) (2009) Herausforderungen im deutschen Luxusmarkt. Köln/Berlin

Guldager S (2016) Irreplaceable Luxury Garments. In: Gardetti M-A, Muthu S (ed) Handbook of Sustainable Luxury Textiles and Fashion, 2nd edn. Springer, Singapore, p 73–98

Meffert H, Burmann C, Kirchgeorg M (2012) Markteting. 11 Edn. Gabler, Wiesbaden

Merck J, Fleischer C (2013) Strategische Positionierung Durch Nachhaltigkeit. In: Riekhof H-C (ed) Retail Business, 3rd edn. Springer, Wiesbaden p. 115–130

Petersen S (2015) How sustainable luxury can save the planet. In: The Huffington Post, http://www.huffingtonpost.com/soren-petersen/how-sustainable-luxury-ca_b_6118512.html. Accessed 30 Nov 2016

Pinkhasov M, (2015) From sustainable luxury to luxorious sustainability. In: The Huffington Post, http://www.huffingtonpost.com/misha-pinkhasov/from-sustainable-luxury-t_b_7069074.html. Accessed 30 Nov 2016

Sauers, J. (2010) Is "Sustainable Luxury" A Contradiction-In-Terms? www.jezebel.com/5649145/is-sustainable-luxury-a-contradiction+in+terms. Accessed 30 November 2016

Stella McCartney (2016) Nachhaltigkeit. http://www.stellamccartney.de. Accessed 28 Nov 2016

Valentino (2016) Cooperate Information. http://www.valentino.com. Accessed 29 Nov 2016

Van Looveren Y. (2016) Stella McCartney puts a price tag on its ecological footprint. In: RetailDetail. https://www.retaildetail.eu/en/news/mode/stella-mccartney-puts-price-tag-its-ecological-footprint. Accessed 3rd December 2016

Qian C. (2016) Why we're mapping China's hazardous chemicals facilities. Greenpeace (ed). http://www.greenpeace.org/eastasia/news/blog/why-were-mapping-chinas-hazardous-chemicals-f/blog/57555/. Accessed 29 Nov 2016

Integrating Sustainable Strategies in Fashion Design by Detox 2020 Plan—Case Studies from Different Brands

Tarun Grover

Abstract In this chapter, our research has been focused on a campaign to overcome the people from the slavement of hazardous chemicals for benefits of short intervals. Different Brands have been recognized with their opinions on the look of their detox 2020 deadline to amputate deadly chemicals not only from their products but from the entire supply chain, if they have the necessary tools to be fit for 2020. For decades, industrial companies have become the major culprit to use the environment and in particular our waterways as a deposition ground for hazardous chemicals and unaffected policies by inefficient government provisions. Water pollution has become a nightmare for local communities that living near manufacturing facilities. There is no 'safe' level left particularly in the Global South, because for persistent, hazardous chemicals and loose provisions have not always clogged the emission of poison chemicals into the atmosphere. In future, government will be pushing for exuberant exemplary alteration on dynamic towards "terminating and unfastening the loop". During the campaign, it appeared that there are two problems which are encountered on the progression of sustainability. First of all, it illustrated out that many fashion companies still have little transparency on their accountability of supply chain, especially the transparency down to farmer level is lacking. Hence, it is very difficult to source materials without transparency in a responsible way. In addition to the speed of the fashion seasons is already causing high pressure for designers working in fashion companies as some companies produce, dump and throw up to 12 collection in a year. People in the industry do not expect this to change because it responds to the demands of the consumer. As an outcomes of the research recommendations on sustainable design strategies are put together in a final product, targeted at textile companies, fashion companies and fashion designers. The final product exists after following the mandatory guidelines from a manual for sustainable fashion design, and this handbook could serve as a helpful tool for the implementation of (one of the) strategies.

T. Grover (✉)
Department of Design, Manipal University, Manipal, India
e-mail: tarun.grover2010@gmail.com

© Springer Nature Singapore Pte Ltd. 2018
S.S. Muthu (ed.), *Detox Fashion*, Textile Science and Clothing Technology,
DOI 10.1007/978-981-10-4783-1_3

Keywords Sustainable · Detox · Closed loop · Slavement and hazardous chemicals

1 Sustainability

Sustainability is an alarm that is used to protect the sphere in upcoming future. The impact of hazardous chemicals are spreading in fulfilling the desire of designers to meet their targets that affected very badly on atmosphere. This has concluded the serious notice of policy makers and organizations for environment friendly development. World Commission on Environment and Development (WCED) directed its appropriate definition on sustainability is '**meet the requirements of the present without bargaining the capability of future generation to meet their requirements and aspirations**' (Fig. 1).

Often, Sustainability is nothing but relying on four major parameters like culture, economy, society, and environment. We are disposing off the existing materials just for sake of to fulfil the desires and demands for the new products, new markets which are leading to wastes, landfills and extra materials for our future generation.

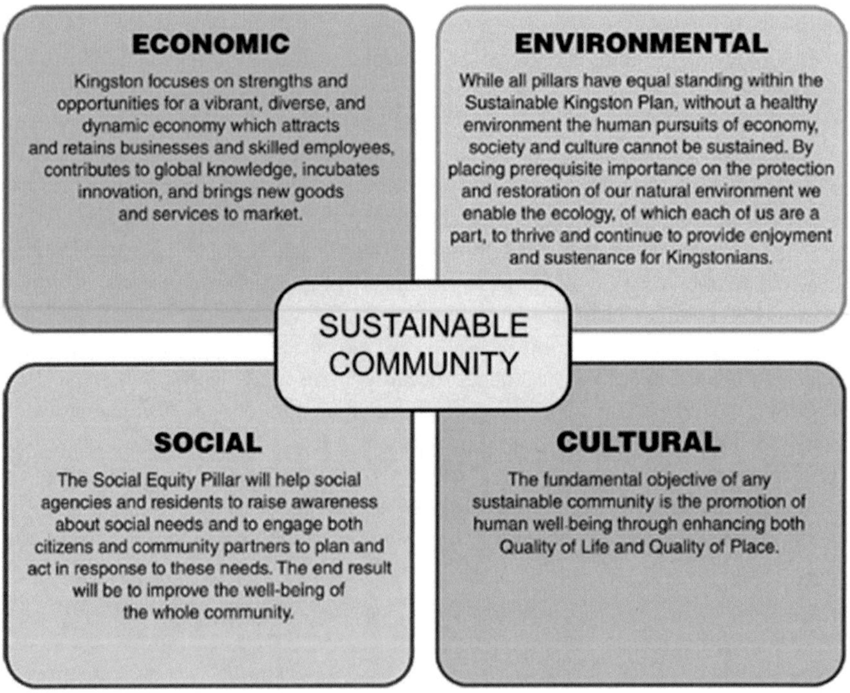

ECONOMIC

Kingston focuses on strengths and opportunities for a vibrant, diverse, and dynamic economy which attracts and retains businesses and skilled employees, contributes to global knowledge, incubates innovation, and brings new goods and services to market.

ENVIRONMENTAL

While all pillars have equal standing within the Sustainable Kingston Plan, without a healthy environment the human pursuits of economy, society and culture cannot be sustained. By placing prerequisite importance on the protection and restoration of our natural environment we enable the ecology, of which each of us are a part, to thrive and continue to provide enjoyment and sustenance for Kingstonians.

SUSTAINABLE COMMUNITY

SOCIAL

The Social Equity Pillar will help social agencies and residents to raise awareness about social needs and to engage both citizens and community partners to plan and act in response to these needs. The end result will be to improve the well-being of the whole community.

CULTURAL

The fundamental objective of any sustainable community is the promotion of human well-being through enhancing both Quality of Life and Quality of Place.

Fig. 1 Sustainability concept. (https://www.cityofkingston.ca/residents/environment-sustainability/sustainability)

Constantly, it has becoming a serious threat for our environment; because we have to protect the sphere in future. In order to conquer and resolve these intentions, we require effective and efficient headship blueprints along with estate to execute transparently.

During the past years 'sustainability' has almost become an overused term in the fashion industry as well as in other sectors. There are many books, articles and conferences continuing the discussion on the meaning of this term. In fact this term is covering many aspects and understandings, often relying on its context. To most people the key principle of sustainability is to preserve the earth and its resources for the next generations, but for every business or organization sustainability requires a different approach. The term 'sustainability' was described by Fronteer Strategy as a moving target, determined by an organization and its ambition, market, environment and its stakeholders (Fronteer Strategy & Green Inc 2012).

Nevertheless, the question is where this sudden interest and engagement of sustainability comes from. It has been suggested by critics that sustainability is today's new ideology. The need for an ideology is inherited in the human nature, humans need an ideal to live for. As suggested by Balakrishnan et al. *sustainability* is 'merely a function of market forces, which will generate the solutions for all problems including the environmental dilemmas that loom over the globe today (Balakrishnan et al. 2003). This suggests we want to believe that sustainability will solve all our problems arising (ecological crisis) from our consumption. This suggestion is being supported by the philosopher Zizek who claims that 'sustainability is the new ideology for addressing a problem'. This could be a logical approach to the sudden popularity of this term, according to the everlasting need for ideologies by humans what is also called 'Ideology mechanism', caused by the temptation of meaning (Zizek 2010). One could argue that sustainability is a new ideology, however, by others it is referred to as a very serious subject saying "Few conversations hold greater portent for the future." (Ehrenfeld 2008). Ehrenfeld cites the following definition of sustainability: "the possibility that human and other life will flourish on the planet forever", suggesting that sustainability is as old as the emergence of human cultures and that it provided a cultural basis for the emergence of magic and religion which serve "to illuminate sustainability and to seek it as part of one's living experience" (Ehrenfeld 2008). with his book 'Sustainability by Design' he wants to point a way to a sustainable future by providing new strategies on design in different sectors. Even in the conducted interviews philosophical thoughts on sustainability were shared. One of the interviewees sees sustainability "something we look for since our quality of life is already very good".[1] Hereby referring to sustainability as a new kind of ideology, just like another interviewee who suggested that probably the most sustainable humans are the poor because they do not consume what they do not really need. Overconsumption is the cause of the exploitation of resources and a one hundred percent sustainable products do not exist because this means there would be no product at all. (Charter and Tischner

---

[1]Appendix D.

Fig. 2 Sustainable art form
derived from nature
[self-made]

2001). this research aims at a more sustainable future by design in the fashion industry. However, only several aspects can be addressed in this research since sustainability is a very broad definition. The sustainable design strategies analysed in this research were selected on influencing the following elements:

- Waste attenuation;
- Cleaner manufacturing;
- Cleaner articles;
- Less materials;
- Less energy;
- Renewable resources;
- Renewable energy;
- Reduplicating and Reclaiming;

With these aspects in mind a definition was created for the term sustainability. When regarding to sustainability in this research it is defined as: **The (re-)use of a minimum of resources like water, energy and fibres, the reduction of waste material and the extension of a garment lifecycle** (Fig. 2).

1.1 Role of Sustainability in Textile

Sustainability performances a crucial role in textile processing right from the beginning of pre-treatments (de-sizing to mercerising) to coloration of textile substrates. To serve the environment eco-friendly as well as customer preferences has become major concerned along with have ability of a company to support its economy should not fall flat. Let's take an example of cotton T Shirt; thousands of gallons of water are consumed in its pre-treatments from unfinished to finished goods on the niche markets i.e. For an instance primarily step should be governed in the growth of cotton in which various chemicals are sprayed to protect from different types of harm full pesticides. Hence, we should cultivate the organically produce cotton which can be vegetated in less water without using chemicals and harm full materials. Henceforth we can save the exercise on pouring of plenty water and harm full chemicals. In developing countries, sustainability is not a priority that's why all manufacturing and production capacities have been a major role in spoiling the environment by consuming endless packaging materials and energy resources. Moreover, we cannot foist on this responsibility on company management and government but it is the accountability of individuals who use these products and then dispose off it for their own interest. Furthermore sustainability is not only a concept but an approach that must aspire to become the norm—not the exception. The entire textile sector has realized the demand of organic products over synthetics desires which mainly influences of their spending behaviour.

1.2 Role of Sustainability in Fashion Industry

Fashion has huge impact on sustainability by pulling and pushing the different categories of fashion like classic, fade and punk etc. Its vicious cycle of manufacturing and designing often depends upon faces of celebrity who does launch in particular markets. If fashion designers are little conscious about dramatic impacts of hazardous chemicals and social concerns about the selection of raw materials, consumption, supply chain and disposal of fashion products, they could diverge the population towards sustainability. Although, Being well dressed is not only a solution to save the planet from present and future sustainability of clothing and textiles. This chapter suggests about the significant and major impacts bearing the fashion industry were primarily to know the current situation of particular country in regard of available resources and their environmental impacts. The main environmental problems allied with the fashion industry are:

- Strengthen fibre production, and laundering clothes;
- proper selection of raw materials to fulfil the supply chain;
- the use of toxic chemicals, which are harmful to the environment and human health;
- the release of chemicals in water systems;

- Fashion and Textile waste as an outcome of fabrication processes and disposal of products.

Awareness about safety and social concerns for Industry:

- Abandon the urge of child labor; abuse of a low or unskilled workforce;
- Take care of their employees conditions in pay and employment;
- Sexual harassment.
- Training about the health concerns associated with the industry included: Hazardous chemicals.

The designer's role becomes magnificent in order to adopt sustainable strategies within the fashion design process but it is not easy as it said earlier. Because fashion is act as a bullet fired from a gun like every person demands high fashion, exclusive outfits that no one has ever seen it. Therefore, it is not the responsibility of fashion designer who is creator of those products but everyone takes as a moral responsibilities to avoid the shadow of chemical treated fabrics and garments. Apart from these, people should use their wardrobe as maximum it possible so that we can fulfil the supply of products made from sustainable practices. However, it can be concluded that the fashion designer is in a position to address some of the environmental and social issues that are associated with the fashion design and production process. Furthermore, In order to promote the production of eco-friendly products, Govt. should interfere and formulate some policies and give some rewards to fashion designer so that they can easily opt it under accepted norms.

1.2.1 Implementing Sustainability a Barrier Pro Fashion Model?

Quick response and Seasonable goods are the main accused that are affected by all sectors of the fashion industry. Gradually, Fast fashion is overloaded to meet the demands of globalised clients for sake of to earn huge profits and innovation in this model. From inner wear to outer wear, high street brands to luxury labels, need to turn up into sustainable products and services, which consider environmental, economic, social and ethical issues through and through the product life cycle. As a society the obsession for consuming fashion goods has seen an enormous growth in the 'fast fashion' sector. As it name implies this sector is responsible for the pace of production and trend driven products that utilize 'at nick of time' technology. In order to reach the products at defined destinations has multiplied its speed by internet and technological innovations. Pricing is one of the factor in this competitive sector that focus on selling fashion garments that retail at a low price who decides the potential of products. Characteristically, Designing is done in one country, sourcing is from other country and production is accumulated in the country where they get cheap labour. In order to fulfil the complete cycle, lots of problems generally encountered like logistic handling is one of pressure that faces by workers of an industry. This whole concept gave a birth of harassment with cheap labours through outsourcing or contracting the labours where low salaries and poor working conditions make it more worsen.

In order to fulfil the demands of fashion, it has been streamlined into two categories like 'quick, cheap fashion fix' in which the garments are recognized as short lifespan products by the consumer and Expensive/exclusive fashion mix is recognized by elite members of society in which they use at once and sell it into markets. In general by hiking the salaries of employees, they have accepted that fashion garments are disposable items since it diversifying in categories of products. For elite group of people to reveal that even exclusive clothing are often disposed off and replaced rather than renovated. However, the fashion garment is almost inevitably recognized as an item designed for a particular 'moment in time'. Designing is the stage where whole sustainable picture depends, because it accounts almost 60–80% of life cycle impacts of a product are created at the design phase. According to UNEP 2004, fashion designers need to produce sustainable fashion garments by changing in their own way that they generally do the things. It follows that in order to stimulate design students and designers in industry to reflect upon their design practice and integrate sustainable strategies in the fashion design process, it is essential to counter the present approach of fashion designing.

2　Materials

The fashion industry is meant to insist its customers to abandon the chronic items (that are still usable) they are utilizing and purchase the new and more adorable products. In order to emphasize on reverse design process, Industry should try to reclaim and reconstruct the old garments for sustainable development (Hochswender 1990). Materials play a significant role in making a path of sustainable textiles and fashion products that not only well-looking but also captivating. Fast fashion is a central business in textile and clothing industry. Fashion is still leading among number of styles and design that caters for the fundamental human needs like protection and modesty and simultaneously fulfil the request for ornamentation and beautification. Although, footprints of fashionable textiles reveals in human culture that it is the forefront of both artistic and technological improvement (Fletcher 2008).

To producing fashionable textiles, it has been divided into two different types of fibres: natural and synthetic fibres. Natural fibres are extracted from plant resources and different animals while manufactured fibres are synthesized in laboratory by synthesizing of man-made materials and sometimes, raw materials also derived from plant and animal resources. The quest of cotton or natural fibres manufacturing for organic garments are increasing all over the world, but unfortunately today the industry is administering majorly for few and similar raw materials. In textiles, the global consumption all over the world are cotton and polyester fibres; these blended fibres are used in large scale application and grant about 75% of the

Table 1 Fibre types (Fletcher et al. 2008)

Natural fibres		Manufactured fibres	
Plant resources	Animal resources	Synthetic polymers	Natural resource
Cotton	Silk	Polyester	Rayon
Hemp	Wool	Acrylic	Modal
Jute	Cashmere	Nylon	Lyocell
Flax	Mohair	Olefin	Carbon fibers
Ramie	Qiviut	PVC	Acetate
Bamboo			
Sisal			
Banana			

worldwide market in fashion and textiles. In last 15 years, demand for polyester fibres has been doubled and made it most prevailing and dominated textile material after cotton. This influenced in largely production of limited fibres in a specific agricultural sector that reduces the consumer preferences and amplifies many risks like ecological and environmental risks (Fletcher 2008).

Few industrialist have been thriving to grow (low chemical) cotton organically like flax and hemp that urge less water and pesticides for their structure. We should stress out on producing new type of fibres that are more cost efficient. Similarly, we should get rid of the usage of polyester which reduce our subordination on oil for polyester production. We can go for biodegradable and renewable fibres such as Alginate fibres and Bast fibres, as these fibres have low impact on environment. Sustainability can bring diversity by selecting the proper raw materials and efficient method of cultivation of bio-degradable fibres which results less resource consumption, more advantage for local farmers, more employments and eventually more social, economic and environmental benefits and will promote local agriculture with effective techniques (Table 1).

2.1 Significance of Materials

Materials play a vital role in manufacturing any type of garments where our mother earth can provide these materials to us. These materials of clothing distinguishes in the status of people in society like if fabric is more stiff and more consumed chemicals dominates a well off human being in society and it also brings out our opinions about economic, environmental and social responsibilities, because there are many crucial steps are puzzled in the production of natural fibres and man-made fibres. The transformation of raw material into a finished fabric leads to several stages which involves labor, energy, water and chemicals etc. Therefore, it is very important in the production of a fashion product that demands attention relating to environment, quality control, and sustainability from agriculturally produced fibres.

2.2 Improving Conventional Fibres

2.2.1 Organic Cotton

Choosing organic cotton over conventional brings several benefits covering different aspects including: the use of less chemicals, reduced water usage, no use of genetically modified cotton crops and the use of low-impact dyes. Although one might expect organic cotton to be cheaper than conventional cotton as pesticides and fertilizers account for 50% of the cost price, unfortunately this is not the case. When growing organic cotton the problem of weeds and insects is something what needs to be dealt with, and doing this in an organic way costs even more than the use of fertilizers and pesticides; this is why conventional production methods use these chemicals. The harvesting of organic cotton is more labor intensive than conventional cotton. At the moment the cost price of organic cotton is still 10-45% higher than conventional cotton (Everman 2009). However, according to chapter two, consumer are willing to pay up to 28% more for sustainable products. In fact this price difference could partly be covered by this alacrity of the consumer, in case the sustainability aspect is communicated well and transparency increases. According to a report by the organic trade association it shows that the production of organic cotton grew by 20% over 2007–2008, this is showing a bright perspective for the future (Organic Trade Association 2009). Nevertheless organic cotton needs more land to be grown and for farmers who want to switch to organic cotton production it takes up to three years to become officially recognized for organic production (Fletcher 2008).

2.2.2 Recycled Textiles

The recycling industry has not innovated for a long period, the same technology for recycling is still being used; machines are tearing apart the yarns while it breaks the fibres and due to the shortened length the yarn is of lower quality at the end of the process (Fletcher 2008). Besides, the recycling process is often a slow and costly process. Despite the fact that few innovations took place, recycling offers a low-impact alternative to other fibres. The most common available recycled fabric is polyester, made from plastic bottles by chemical recycling. Although chemical recycling requires more energy than the extraction of fibres, it does provide a more predictable quality (Fletcher 2010).

2.2.3 New Alternatives Bamboo

Bamboo fabric is made of cellulosic fibres and is hypoallergenic, fast-drying and absorbent which makes it a comfortable fabric to wear. Another positive factor is that the bamboo plant grows very quickly, and it can help to improve soil quality. In spite of this, to be certified as an organic textile bamboo should be harvested without the use of pesticides or fertilizers. The production process of the most commonly produced sort of bamboo textile is very similar to that of viscose/rayon, in which a chemical solution is used during this process in order to spin the fibre. Recently several clothing companies claimed they were selling 'eco-friendly' bamboo while actually this was Rayon made from Bamboo, this is an example of greenwashing (Nordic Initiative, Clean and Ethical 2012). A more sustainable way of processing bamboo is the mechanical way (not chemical) which makes the fibre look and feel more like linen so this is a natural way without using chemicals (Fletcher 2008). However, this process is labour intensive and therefore costly. When it comes to the growth of Bamboo, only China is growing bamboo on a commercial scale and due to the difficulties with transparency on production in China there is often little evidence that the fabric has been produced in a sustainable way. Besides, there is not an official certification yet for bamboo fabric (Carter 2008). A new certification would definitely make a difference, as the companies then have more clarity on what kind of bamboo they are buying.

2.2.4 Hemp

Hemp is one of mankind's earliest fibres and it was one of the most important fibres in the EU, until it was replaced by cotton (Black 2008). Just like the bamboo plant the Hemp plant is a fast growing plant, it grows in almost all conditions and it needs little or no fertilizers and pesticides. Besides, it requires little pesticides to grow it and unlike many crops, it enriches the soil it grows in rather than depleting it (Black 2012). In line with Bamboo, Hemp can only be certified organic when processed as a bast fibre, this means it has processed in the natural way not with chemicals. Today most of the hemp available is grown in China, Romania, Russia and Poland subsidized by the EU (Black 2008).

2.2.5 Lyocell/Tencel

This fibre is associated with two names because Tencel is the branded name of the lyocell fibre. It is made from the pulp of eucalyptus trees (grown at FSC farms), which makes it biodegradable and renewable material (Fletcher 2008). During the processing a non-toxic solvent is used which can be recycled and this makes the processing already a lot more sustainable then other fibres (Organic Trade Association 2009). Another benefit of Lyocell is that it does not need to be bleached as it is already very 'clean', and it can be laundered at low temperatures. However, the production of Lyocell is energy intensive an aspect which is one of the challenges for Lenzing, the producer of Lyocell.

3 Fibre to Consumption

Fibre is the first form of fashion textile products and then it converts into yarns by number of techniques like dry, melt, dry-jet and nano-jet spinning's. Then onwards, it passes on looms for the turning up into fabrics which are commercially available in the market. That fabrics are treated with suitable finishing's to make it softer and supple which results into expensive one. To be in the race of competition many textile firms are doing extensive search on fibre and unfinished materials to build their product lustrous, sense soft, supple, and comfortable for the consumers (Akira 2000).

The extraction of raw material of fibrous goods (Fibre) can be categorized in two types on behalf of their production. The fibre can be natural fibre and man-made fibre. Natural fibres are obtained from natural source like plants, animals and man-made fibres are obtained from oil, petroleum and minerals. Each companies are engaged in research and development activities for making new fibres like flame-retardant, water-repellent in order to compete with fellow companies. The Research and development flourishing drastically especially in textile sector which have contributed in all products that uses by human beings right first from he gets up till his sleep like toothbrush to comfortable bed-linen cushions (Akira 2000). The outcomes of natural fibres are obtained from natural resources like plants, animals and minerals. It can be further classified in different form according to their moulding techniques. Further, Manmade fibres have been categorized in two type stages: Regenerated fibre and synthetic fibres (Fig. 3).

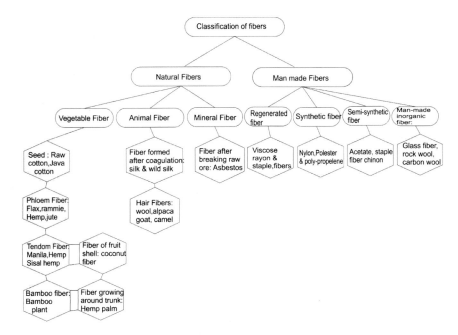

Fig. 3 Classification of fibres (Akira 2000)

4   Sustainable Design Strategies

4.1   Detox Catwalk from Fashion Industries

These companies have taken towards their Detox commitment depending upon the critical steps, the Detox Catwalk groups have divided into three categories: the first two are *avant-garde* and *evolution mode*. Companies have insisted to not deliver sufficient results in a third category: *faux pas*.

This new edition of the Catwalk assesses 19 Detox committed companies from the Fashion and Sportswear sectors. It finds quite interesting results that a few companies are afore of the deadline and on line to meet their future pursuit—these three are *avant-garde*, *evolution mode and faux*. *Evolution mode holds its* majority among twelve of the Detox restricted brands and should give stress in improving their performance in at least two of the three key evaluation criteria. Finally, four brands are unveiled to come across in *faux pas*—by not yet implementing the right away steps needed to achieve the goal of dislodging hazardous chemicals by 2020 and accepting individual responsibility for their hazardous chemical pollution.

- **Avant-Garde**: a good performance on at least two criteria and no critical failure on any
- **Faux Pas**: critical failures on at least two criteria
- **Evolution Mode**: any combination of assessment between Avant-Garde and Faux Pas as described above.

In order to commitment of detox, there are many companies whose products have been found to contain hazardous chemicals and have not well answered with a reliable detox commitment to overcome this issue. This edition assesses the deeper implementation of the companies Detox Commitments—to address the question, are the Detox brands 2020 fit?

By successfully achieving their Detox Commitments, Various companies are taking an important steps to clean up the release of hazardous chemicals throughout the environment and keen on the goods of textile supply chain. However, the increasing environmental concern and globally, health impacts from the manufacture of textiles are not only the major issues but also impacts across the whole textiles life cycle are amplified by the huge increase in the quantities of clothes that are sold then the rate that they are thrown away. Henceforth, this edition focusses, the issue of **"closing and slowing the loop"**—for forthcoming events and Detox campaign will measure the duties of individuals whether brands´ are approaching towards closing and slowing the loop in order to unravel exorbitant urge of instruments/resources.

Criteria

These are three criteria to measure on the issue of hazardous chemicals:

1. Detox 2020 Plan
2. PFC Elimination
3. Transparency.

Through a familiar design, Companies will compare their performances with the best practice in each criteria category. It will be accessed, if companies are aligns on track to meet the goals set in their Commitments to Detox by 2020.

- **Detox 2020 Plan** focusses on a company's chemicals management system, specifically its Manufacturing Restricted Substances List (MRSL) is needed to identify hazardous chemicals used in manufacturing by suppliers and set priorities for elimination with timelines.
- **PFCs Elimination** currently serves as a directional guide for the implementation of the 2020 goal. It assesses the progress made towards the commitment to eliminate any use and discharge of one of the widely used hazardous chemical groups per/poly fluorinated chemicals (PFCs)[2] and the publication of case studies showing how this has been achieved.
- **Transparency** confirms whether the company has ensured it discloses its suppliers list (including second tier where the wet processes are likely to take place) or its vendors regularly publish data on the dislodgement of hazardous chemicals from their wet processes on the Detox section of the IPE online platform (http://www.ipe.org.cn/En/pollution/discharge_detox.aspx).

Progress is to be reported by companies on their corporate website on a dedicated Detox webpage, visible and prominent, and clearly divided by criteria.

4.1.1 Move Towards Organic Cotton Over Traditional Alternatives

Organic farming has been more susceptible than conventional cotton due to a problem of pest management, and cotton is the most demanding fibre that consumed worldwide in different application so to make it grow fast, pesticides are used on their crop but these pesticides have consequence in ruining the qualities of cotton fibre like absorbency. Organic cotton are more preferred over conventional system where pests are managed by their natural enemies (Fig. 4).

The above figure represents a methodical outlook of all procedures and foots involved in the development of an organic cotton that had been systematically

---

[2]PFCs refer here to per- and polyfluorinated chemicals (also known as per- and polyfluoroalkyl substances, PFASs. It includes precursor chemicals such as fluorotelomers that can degrade to form perfluorinated chemicals (e.g. PFOA), and covers both non-polymeric & polymeric chemicals. See https://www.oecd.org/chemicalsafety/riskmanagement/Working%20Towards%20a%20Global%20Emission%20Inventory%20of%20PFASS.pdf.

A globalized approach for evolution of biological cotton

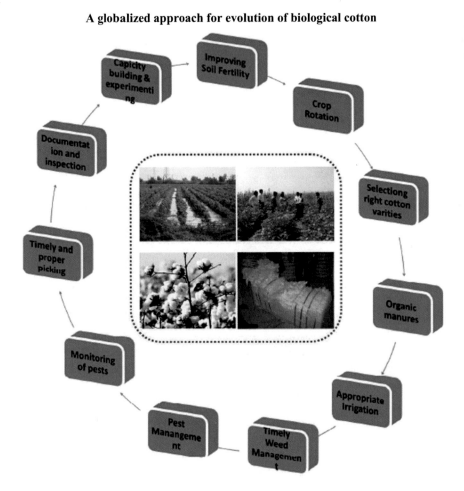

Fig. 4 Presenting a methodical approach for evolution of organic cotton. *Source* (FiBL u.d.)

described above. The organic cotton has streamlined its effectiveness by not using any type of pesticides to cultivate and grow cotton in natural suffice. The organic cotton is an evolving solution in these hostile circumstances. It gives benefits to companies who are investing in organic culture by adopting the scientific concept "making more with less" in many ways; it stimulates up the financial condition of farmers by adopting local infrastructures and varieties, it creates healthy environment and it motivates socially by giving more benefits to the farmers and cut down the risks of toxic chemical impacts on its surroundings by eliminating the procedures of pouring fertilizers and pesticides until its cultivating stages. Until, its pre-treatment steps, the toxic chemicals (Formaldehyde, Aromatic solvents, Chlorine bleach, Azodyes, etc.) are not urged from spinning, weaving, finishing and dyeing (Sanfilippo 2007). The growing of an organic cotton has a multi-directional

Table 2 Differences between organic cotton and conventional cotton

	Convential cotton	Organic cotton
Environment	Water pollution	Improved water utilization
	Loss of bio-diversity	Improved bio-diversity
	Adverse changes in water balance	Soil and air are hygienic
	Pesticides killing beneficial insects	Eco balance between pests and insects
Social	Health problems are there where regulatory system are weak	Use of local varieties and resources
	Poisoning and casualties due to extensive use of pesticides	Helpful for low income families due to more premium
Economy	Resource consuming	Less resource consumption
	High production costs	Low production costs
	No alternative crops	Niche markets
Food	Pesticides entering into human food through cottonseed oil	No danger of contaminated edible items originated from cotton source
	Contamination of meat and milk from animal fed to cotton source	
Agricultural	Reduced soil fertility	Improved soil fertility
	Poor irrigation, contamination fields becoming barren	Crop rotation maintains soil structure
Chemicals	Chemicals remained in final products cause health problems like chronic diseases cancer etc.	No use of pesticides that saves the farmer from chronic diseases

Source (Stolton 1999)

effects and favours for our health care and food protection by provisioning toxic set food (Table 2).

Advantages of an organic cotton in comparison to the conventional cotton are abbreviated below.

4.1.2 PFCs Elimination from Entire Products

The categorization of hazardous chemical group is essential to prevent its commercialization like per/poly fluorinated chemicals (PFCs)[3] is one of the 11 priority groups identified in the companies Detox Commitment. PFCs are a group of

---

[3]By publishing chemical discharge data via the IPE disclosure platform, a company's suppliers ensure that the data is credible, that it includes the necessary details to identify the individual facility concerned and that it covers at least the 11 groups of priority hazardous chemicals. Because much of the world's textiles production takes place in China, companies must ensure data from suppliers in China (including Taiwan) is disclosed, followed by other major suppliers in the Global South.

chemicals that and have been identified as persistent, bio-accumulative and/or toxic and they are known for their water and oil repellent properties (www. greensciencepolicy.org/madrid-statement). Responsible Detox brands have committed to separate out (and some have already eliminated from their global supply-chains) of its any use in products and discharge of PFCs.

The commitment is to reduce the amount of circulation of PFCs as a **group**, the elimination of following a precautionary approach, rather than chemical by chemical, as chemicals with a similar structure and properties can be expected to have similar properties and hazards. PFCs are particularly relevant for the implementation of the precautionary approach in the EU's REACH regulation, where they are potentially being looked at as a group.

Principles for hazard screening methodology:

1. Hazard based: no 'risk based' criteria for excluding certain chemicals
2. Includes a broad range of hazardous categories (at least those under REACH regulation)
3. Using at least all publically available information
4. Cautious thresholds in hazardous criteria setting
5. Assessment of the effectiveness of the screening tool for hazard identification
6. Full transparency on criteria, methods, data, thresholds, information sources
7. Taking by-products and environmental fate into account
8. Recognize the importance of physical form e.g. nanomaterials, polymers, etc.
9. If no or missing information the 'hazardous until proven non-hazardous' assumption should apply + group approach.

4.1.3 Transparency in System

This criteria is more governed to people, the **Right-to-Know**, giving the people the information on risks and impacts and empowering them with a capacity to influence. Being a global citizens, people has some fundamental rights to know about the composition of the products before buying any hazardous chemicals product and the consequences to discharge into the environment.

The Detox promised also includes transparency on the use and dislodgement of all hazardous chemicals to ensure that those responsible answer to public pressure and act effectively and instantly to get along its use and discharge of hazardous chemicals (Greenpeace International 2011) (Fig. 5).

The transparency in the urge of publication of discharge data also seeks to further engage brands with their suppliers and support moves towards a clean factory approach. Publishing data allows all stakeholders to follow, trust and challenge a company's progress and feed self-motivation. It can also significantly influence local legislation to adopt Detox water standards and chemical policy. Ultimately, it could inspire the establishment of a global standard for transparency and accountability across the whole textile sector and beyond.

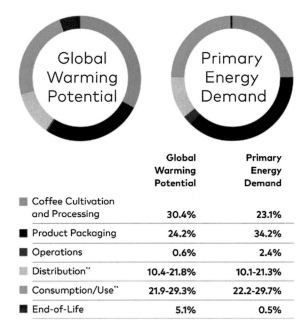

Fig. 5 Case study of one industry operations (http://www.keuriggreenmountain.com/en/sustainability/sustainableproducts/productimpact.aspx)

	Global Warming Potential	Primary Energy Demand
Coffee Cultivation and Processing	30.4%	23.1%
Product Packaging	24.2%	34.2%
Operations	0.6%	2.4%
Distribution"	10.4-21.8%	10.1-21.3%
Consumption/Use"	21.9-29.3%	22.2-29.7%
End-of-Life	5.1%	0.5%

4.1.4 Engagement of Designers to Opt Sustainable Practices: Through Design Process

> The design process within Fashion Company can also be understood as 'Collection Development'. Designers working for a fashion company are not always completely free to design anything they want, they are bound to many rules and regulations and often working under high (commercial) pressure. These restrains are for example a certain amount of product groups and styles, trends in fabrics or styles and of course there are financial boundaries as well, all set by the planning department. Next to this, 'the designers' ability to check up on their suppliers is limited by the very tight time schedules of seasonal fashion production, which are pivotal for commercial success (Black and Eckert 2010).

These factors are restraining the possibilities for designers to create more sustainable products. The influence of the designer on this process is also determined by the certain role the designer has in a fashion company, as this differs per company. These roles have been described in a theory by Bakker. He suggested that there are two roles a designer can have in a fashion company: an operational role or a strategic role. In the operational role the designer works with an idea that he or she translates in a product, taking in account the constraints of the 'company rules'. A designer in an operational role has little influence on the environmental influence as the designer has modest control over the fundamental product idea. The second role is the strategic role. In this role the designer is involved in the product planning stage, having a say in strategic decisions on the product like: function, physical properties and environmental profile (Fletcher 2010). The processes before the actual design stage are visible in Fig. 2: many researches and analysis are conducted after which a product planning will be made (Eberle et al. 2004). According to these plans a designer starts the design process. Bakker added that most designers' roles in fashion companies are more likely to be operational. From this viewpoint we can already conclude that designers often do not have the ultimate position to be able to influence the environmental profile of a product.

Moreover, the willingness to create more sustainable products is also highly dependent on the interests of the designer. After product planning, sketches are made by the designers and together with fabric samples they are discussed. After the first approval the samples are ready to be made, either at the company itself but often by outsourced production. When the company receives the sample (a sample will be received for almost every style of the new season, except for styles which are repeated), these samples are fitted and judged on quality. Based on these fitting and quality check sessions, samples are approved on and bulk production can be started. The ones which have not been approved, will be changed before going into production or will be cancelled. After the bulk has been produced it will be shipped to the stores. According to the conducted interviews with professionals who are working in the fashion industry, it appeared that for many designers and other professionals in the fashion industry transparency is lacking. As the production is often being outsourced it is hard to control or to influence what is happening at the place of production unless one is travelling there. Furthermore all the interviewees agree that the speed of the collections in fashion is very high, often bringing up the example of Zara a fashion company designs and produces styles within 6–8 weeks. It is predominantly the demand of the consumer that makes a fashion company increase the amount of the collections a year and makes the fashion company deliver new styles to the store all year round.

4.1.5 Design for Longevity

The market strategies of fashion companies are often developed by the marketing department. This department creates value by presenting the brand as selling quality products and therefore making profits. Through this marketing tool the fashion company reaches its customers and provides them with a brand experience (Nielsen 2010). At the same time, this focus on marketing and market strategies could abstract from the focus on product design, development and manufacturing and therefore the quality of a product. The product price and quality of the products of these fashion companies are often calculated on beforehand in order to meet with the target group and market strategy. In order to become a more sustainable company, the design process should become more involved with this strategy planning. When designers are able to expand the potential quality of products, the product will get more sentimental value and therefore provide a longer-lasting utilization and emotional product attachment by the consumer and even reutilization (Nielsen 2010). A product which is useful and meaningful to a consumer on the long term, will not be easily replaced. A fashion chain striving for more sustainable products should allow more time and budget for research, discussion and debate, and this will enhance the possibilities for designers to make a difference in creating more sustainable products.

In this chapter some practical opportunities for sustainable fashion design will be discussed, according to a theory from Van Nes. She worked from the viewpoint of understanding replacement behaviour. The replacement behaviour of consumers depends on several product characteristics. According to this research she developed a model 'Design strategies for longevity' (Van Nes 2010). This research was based on several types of products not only clothing, however, the model can be applied to garments as well. The details of this design strategy below are slightly adjusted according to similar models of Van Hemel and Charter and Tischner to make it more comprehensible from a fashion industry perspective. The model exists of five categories and sub-categories, of which each could be explained in view of fashion design.

1. Design for wear-resistant design

Reliable and wear-resistant design stands for a level of quality the consumer should be able to rely on, this involves preventing wear and tear. Sustainable garments should have an increased 'Long life guarantee' instead of the low quality garments which are disposed of quickly. Consumers can use their products for a long duration instead of discarding them too soon and buying new (low quality) garments. This will save fibres, energy and water used during production.

2. Design for (easy) repair and maintenance

According to an old saying: 'A stitch in a time saves nine'. This relates to the fact that repair prolongs the life of a garment and this saves nine times the energy what is needed for the replacement of new materials. Due to the fact that clothing has become cheaper consumers are discarding their garments easier as repair costs are relatively high. Besides, consumers rather buy a new item which is 'on trend' and satisfying their regular shopping habits (Allwood et al. 2006). However repair and maintenance could expand the lifecycle of a garment, as many garments are disposed of even before repair. Often, garments are sold provided with a spare button, but to increase the repair of clothing, more inventions are necessary. For example garments could be designed to facilitate repair with easy-to-remove parts (e.g. collars and cuffs on men's shirts) which makes it easy to replace these parts (Allwood et al. 2006).

In order to facilitate this repair spare parts should be offered by fashion companies through, for example, their website. These parts could be buttons, zippers, other haberdasheries, spare parts of garments etc. This also gives an opportunity for consumers to customize their clothing (Design For upgradeability).

3. Design for upgradability

Design for upgradability are products designed with the consumption of a product in mind. It 'enables the opportunity to add new functionality (and fashion appeal) during the life of a product by replacing parts or modules (Van Nes 2010). Therefore parts should be easily replaceable, so that the consumer can decide to give the garment a new look and/or functionality and therefore expand its lifecycle.

This design strategy is slightly different than the second strategy because it's more focused on the need for change of the consumer rather than necessarily replacing parts due to wear and tear.

4. Design for product attachment

This design strategy can be related to that of co-creation, as it is related to personification of a product at the acquisition process. Besides, Design for product attachment enables the consumer to 'customize' during the consumption process, for example, adding personal elements to the product, or ageing with dignity. The last method has a large potential in the fashion industry, especially concerning the consumption of jeans. The production of jeans with a worn look requires treatment which are non-environmental friendly (e.g. bleaching, stone wash, sandblasting, scraping etc.) and already decreases the quality of the fabric. Unwashed jeans are stronger and has a longer durability, but requires a much longer period for the wearer to create this worn look (LG 2012). However when wearing unwashed jeans, the jeans will get authentic and personal characteristics and therefore more emotional value for the consumer and a longer lifecycle.

5. Design for variability

A design for variability can be easily changed by the consumer, example a jacket or pants which can be worn inside out. The consumer can convert the look of garment into a different one without having to do a lot of effort. According to a research by Koo 'Changeable garments have the capable to lead customers' natural business with sustainable deeds by fulfilling their various requirements and wants even though consumers may lack of cognition or shows little importance about sustainability' (Koo 2012). Expandable garments are being worn for a longer period of time and more frequently due to serving multiple needs of the consumer. Koo's research gives an insight in designing transferable garments. Design for longevity is a strategy that requires dynamic, flexible products. Therefore the designer will need to think ahead of what is going to happen with the product during its lifespan. As a result variability, product attachment and future repair or upgrading can be implied (Van Nes 2010).

6. Design for disassembly

A large part of garments exist from blended textiles, for example a blend of natural and synthetic fibres. This method was named 'monstrous hybrid' by McDonough and Braungart since it is almost impossible to use either one of the fibres for re-use or recycling. To improve this system they suggested the concept of 'design for dismantling, a strategy for aiding material recovery, reuse, recycling of composting' (McDonough and Braungart 2002). When designing for disassembly, the designer should ensure that at the end of a products life the used materials can be reclaimed for reuse. Therefore products should be designed with the simplest forms possible.[4]

---

[4]Blackburn, Sustainable textiles: Lifecycle and environmental impact, New Delhi (Woodhead Publishing Limited).

Fig. 6 Disassembly of garment components (Gam et al. 2010)

This is the crucial element of this strategy, as product materials only have a significant recycled value only when it is divided into clean, separate elements. Mixed materials are impossible to separate in an ecological as well as an economical way (see footnote 4). Over the years, interest for design for dismantling has been raising up, however most research and implementation yet took place in the automotive industry (Gam et al. 2010). During the research of Gam et al. it was found that garments (in this research mainly men's jackets) which are conventionally produced were difficult to disassemble due to the use of several types of textile. For successful design for disassembly three critical factors should be taken in consideration; selection and use of materials, sequel architecture and the selection and use of joints, fasteners and connectors (Gam et al. 2010) (Fig. 6).

Gam et al. developed different suggestions to minimize the time and effort required for disassembling garments and maximizes the material recovery:

1. Aim to bring down material variation, and to stitch parallel materials together during the production of a garment, this will decrease the disassembly steps and time.
2. When sewing different types of textile together (natural and synthetic), a larger stitch type should be used, for example a general needlework with six stitches per inch.
3. In case of the use of fusible interlining (e.g. production of jackets), blind hemming stitches under the collar and on the backside of the lapel (Gam et al. 2010).

These measures applied during production of a casual men's jacket eventually made the dismantling of the men's jacket 1.5 min faster. This could have the same effect when applied in the production of different kinds of garments. When the disassembly process becomes faster, and at the same time less costly, it will be more appealing for fashion companies to recycle textiles in the future. A brand that already involved Design for Disassembly in its business model is Timberland, they applied it on one of their footwear lines. This footwear line is the so called Earth-keepers line, and the styles are created from recycled material. Next to this, the shoes have a greater potential to be recycled after the consumption process. Because of the design for disassembly, Timberland states that 50% of the shoe can be recycled compared to other styles (Timberland 2012) (Fig. 7).

Fig. 7 Recycling after dismantle of shoes components

5   Perspectives from the Professional Field Who Has Adopted Detoxification

To start with, all interviewees working in the fashion industry agree on the fact that a change is taking place in the fashion industry, "the days of usual production and sales of clothing are over." When companies will not respond to this change, "the future of a fashion company is in jeopardy." Sustainability is increasingly becoming important in the policies of fashion companies. When asking the interviewees on the transparency in the supply chain of the fashion industry, a problem which is inevitably connected to sustainability, they agree that it is very difficult (especially for the large companies) to get full transparency of their supply chain all the way to the 'farmer level'.

Another issue in the design and production of clothing is that the speed of seasons and collection drops increased tremendously. The demand of consumers for new styles and trends is forcing fashion companies to produce many collections a year, and sometimes this results in cheap and therefore low-quality garments. Almost all of the interviewees directly refer to the example of Zara, a (vertically integrated) fashion company that is able to design and produce clothing and 6–8 weeks. The pressure of producing many and cheap collections involves all kinds of fashion companies from small to large sized companies. The interviewees do not expect this demand to change very soon as the "consumer is impatient" and the consumers need to change their buying and consumption behaviour into a more holistic approach.

Sustainability could provide a solution to overconsumption, and some interviewees were referring to the idea that sustainability is 'the new ideology'. "Now that people have a very good quality of life, and therefore they start looking around at what their consumption is actually doing to the earth, for example how are clothes are actually being produced." Another interviewee says that "A hundred percent sustainable company cannot be achieved as this means that there should be no products at all". To not produce any products at all is not an option as this would mean the end of many fashion companies. Nevertheless, designing garments that have less impact on the environment is a possibility and it could offer many benefits for a company as well. Therefore design strategies were analysed in this research and through conducting interviews these were critically reviewed.

According to the interviews the first strategy, the use of a wider variety of textiles or new and more sustainable textiles, is already being implemented in small amounts within many fashion companies. Some companies started to work with organic cotton or recycled polyester already. However, it has not been applied yet on a large scale yet as the availability is still little and therefore it is expensive as well. Still almost all the interviewees foresee a positive future for this strategy as they expect that the prices of sustainable textiles will decrease in the future, and it is a strategy that can be implemented on the short term. Currently "the look and feel of the fabric is more important than if it is organic or not" to many fashion companies, but when the offer of organic and sustainable textiles increases a wider variety and larger amounts of fabrics will be available as well. Very helpful for companies which want to source more sustainable fabrics are organizations like the Better Cotton Initiative in which many brands are engaging. With the support of brands, this organization has the possibility of 'chasing' the supply chain and to improve the current operations.

The opinions of the interviewees on the second strategy, co-creation, differ. Co-creation has "the great potential to increase the emotional bond of the consumers with the product and brand." Next to this is expected that "it could boost sales and it is a great marketing tool." On the other hand, co-creation requires a good online platform which could be expensive and most of all, to make it successful it has to be targeted at the right group of consumers. The target group that could be interested in co-creation is a younger group of people, the interviewees agree. This target group is interested in design in comparison to the 30 + target group which is less interested in this aspect. Some interviewees refer to the successes of Nike by implementing co-creation, saying that they target the exact right group of consumers for co-creation (Fig. 8).

(a)   **(b)**

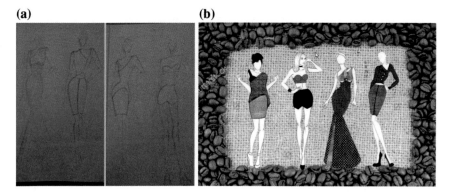

Fig. 8 a Sustainable outfit (Manual sketch). **b** Sustainable outfit (CAD sketch)

Design for longevity relates to a more holistic view on the use of clothing. It has the potential of extending the life of clothing to make it more durable and compared to the current speed of the fashion industry it almost makes it a paradox. Two of the interviewees relate to the electronic industry, which marks the change into a consumerist and throwaway society. Years ago people bought televisions or other electronics and they would have them repaired when broken. Instead, today consumers directly buy a new television because it is relatively 'cheaper' than to have it repaired. It is now actually the same with clothing and no repair service is offered anymore in any store (Horrocks et al. 2007). One of the interviewees suggests that the key to this strategy is to provide designers with information on the consumption of garments. If designers have more knowledge on the consumption of garments, they will be able to anticipate on this in their designs (Sanfilippo 2007). What was also suggested is the use of fabrics that get a different look and feel during the consumption of the garments for example by wearing or washing it (Marchand 2008). Therefore a garment only becomes nice to longer a consumer uses it and therefore it has the potential to extend the lifecycle. The implementation of the last strategy, design for disassembly, depends on the kind of fashion company. It might work for more basic styles or basic fashion companies, but most fashion companies use blended fabrics for the benefits of these fabrics. The strategy might be implemented in the future but at the moment companies are more focused on the use of for example organic cotton as recycled fabrics are expensive and only available in small amounts.

These strategies have are predominantly applicable on large companies as they provide garments on a large scale of which the production has a big impact on the environment. From the interviews it appeared that small companies often have different values than large companies and therefore the strategies are less suitable for small fashion companies (Nielsen 2010).

6   Toxic Level Assessments with Different Brands

The 2017 detox survey found that a few companies are ahead of the mark and maintaining their status to be on track to meet their commitments which are as follows:

1. **AVANT GARDE**—Three companies lies in this grade.
2. **EVOLUTION MODE**—Twelve have committed to be achieved in this level.
3. **FAUX PAS**—Four brands have been come up to achieve this level by not yet accepting individual responsibility for their hazardous chemical pollution and executing the immediate actions needed to obtain the goal of hazardous chemical by 2020.

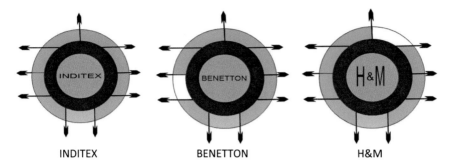

Fig. 9 Detoxy   2020   plan/PFC   Elimination/Transparency   (http://www.greenpeace.org/international/en/campaigns/detox/fashion/detox-catwalk/)

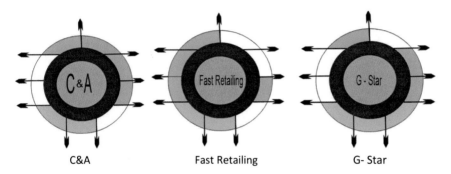

Fig. 10 Detoxy   2020   plan/PFC   Elimination/Transparency   (http://www.greenpeace.org/international/en/campaigns/detox/fashion/detox-catwalk/)

6.1 Outcomes of Brand Analysis

1. **Avant-Guard** (Fig. 9)

2. **Evolution Mode**

Those companies are come under evolution mode are restricted to detox and have made progress in executing their plans, but intervention of concrete actions are needed to evolve faster decisions to get 2020 goal (Figs. 10, 11, 12 and 13).

3. **Faux Pas**

These are those companies who are failed to commit in completely removal of unsustainable practices and currently heading in wrong direction by adopting hazardous chemicals for their supply chains (Fig. 14).

In order to make a credible, individual detox commitments, many of companies who are still not aware about the toxic addicts that have failed to take responsibility for their toxic trail.

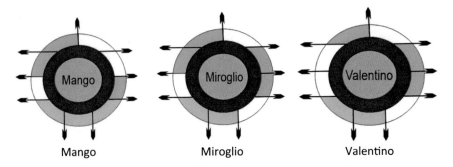

Fig. 11 Detoxy 2020 plan/PFC Elimination/Transparency (http://www.greenpeace.org/international/en/campaigns/detox/fashion/detox-catwalk/)

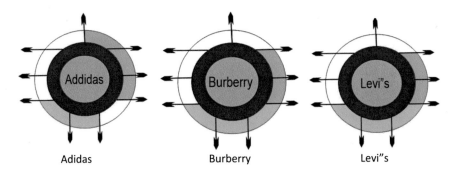

Fig. 12 Detoxy 2020 plan/PFC Elimination/Transparency (http://www.greenpeace.org/international/en/campaigns/detox/fashion/detox-catwalk/)

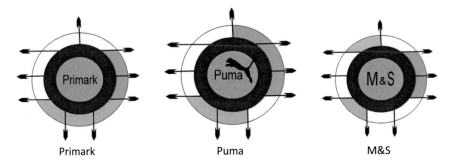

Fig. 13 Detoxy 2020 plan/PFC Elimination/Transparency (http://www.greenpeace.org/international/en/campaigns/detox/fashion/detox-catwalk/)

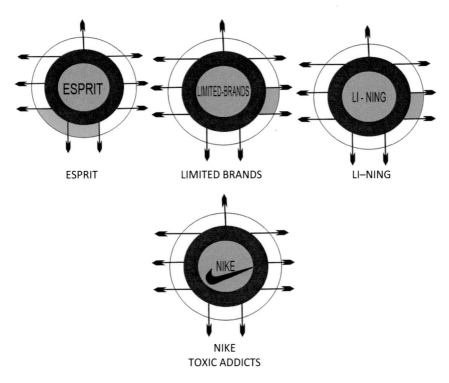

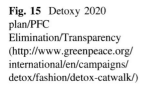

NIKE
TOXIC ADDICTS

Fig. 14 Detoxy 2020 plan/PFC Elimination/Transparency (http://www.greenpeace.org/international/en/campaigns/detox/fashion/detox-catwalk/)

Fig. 15 Detoxy 2020 plan/PFC Elimination/Transparency (http://www.greenpeace.org/international/en/campaigns/detox/fashion/detox-catwalk/)

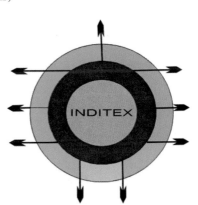

6.2   Discussions on the Outcomes of Brands

1. **INDITEX** (Fig. 15)

- Inditex, is one of the leading brand that commercially produces and sells their products on the name of famous fashion brand Zara, the way its transformed that its tightly action keeps off to using the hazardous chemicals as well as achieving Avant Garde status by delivering it's all benchmarks. Its approach to **transparency** is imitable. It has assured in its publication that they are bound to use only those chemicals which are not hazardous and has published a list of scopex index journals of wet process suppliers.

- This company has well explained its **Detoxy 2020 plan** by using eco-friendly methodology which emphasis on water effluent treatments and hazardous chemicals before it actually disposed off and that reflects a clean clothing approach for the elimination and aspect out of its extensive list of deadly chemicals.
- The timeline of **PFC elimination** indicates well in above figure, that still demands to use broader hazard based criteria to access the suitability of alternatives on PFC's.

2. **BENETTON** (Fig. 16)

Benetton has achieved one step ahead in strictly implementing its **Detoxy 2020 plan** by its performance on transparency in publication of consumed data and meeting all its commitments on eliminating PFC's.

- It has developed a positive **Detoxy 2020 plan**, by executing the principle of clean factory approach that enforce its ban on usage of hazardous chemicals. In order to be consistent, It should be regularly updated with its hazard based screening methodology.
- Benetton has delivered its commitment to eliminate **PFC hazardous chemicals** from its shed and approaching towards a green dimensions.
- It has made a good effort towards its longer term goal on **transparency** by publishing its investigation into its products and supply chain processes.

3. **H&M** (Fig. 17)

This investigation states out that H&M has set an example for other brands to achieve avant garde status by preventive physical waste and pervasive approach in detoxifying its entire supply chain that meets to fulfil the demand of costumers.

Fig. 16 Detoxy 2020 plan/PFC Elimination/Transparency (http://www.greenpeace.org/international/en/campaigns/detox/fashion/detox-catwalk/)

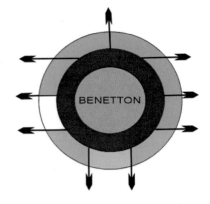

Fig. 17 Detoxy 2020
plan/PFC
Elimination/Transparency
(http://www.greenpeace.org/
international/en/campaigns/
detox/fashion/detox-catwalk/)

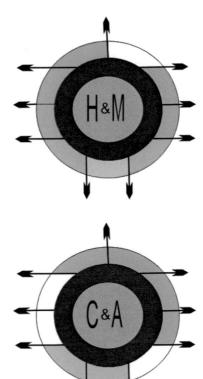

Fig. 18 Detoxy 2020
plan/PFC
Elimination/Transparency
(http://www.greenpeace.org/
international/en/campaigns/
detox/fashion/detox-catwalk/)

- This programme is based on clean clothing approach, hence products of H&M are regularly checked through transparent methods. Regularly updated plans are implemented to regulates its targets and apply detox across their whole mill, not only for brand image/products.
- In order to eliminate all **PFC hazardous chemicals** from its products, H&M was the first one to deliver its commitments on time and publish a case study which helps to expedite the PFC's elimination.
- **Transparency** is an integral part of H&M which aiming the percentage of suppliers reporting their detox data online and helps to fetch its commitments up to date.

4. **C&A** (Fig. 18)

C&A lies in EVOLUTION MODE despite its performance in transparency.

- By referring to its clean factory approach, it has modified its **Detox 2020 plan** by testing its hazardous chemicals and treating its own wastewater treatment. Unfortunately, it still stuck on inadequate screening methodologies which leads to fundamental flaws.

Fig. 19 Detoxy 2020
plan/PFC
Elimination/Transparency
(http://www.greenpeace.org/
international/en/campaigns/
detox/fashion/detox-catwalk/)

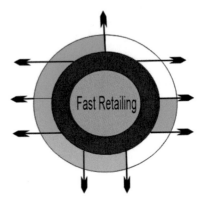

- C&A worked well in recent years to eliminating of deadly **PFC's chemicals** and shares cognition to its suppliers on PFC—free alternatives to use in their products.
- This particular brand worked well in **transparency** level by providing all recipes of PFC free chemicals, waste water investigating report and acknowledging with a list of main vendors or suppliers who provides all manufacturing raw materials.

5. **Fast Retailing** (Fig. 19)

Fast retailing, the company is one step away from avant garde and behind its fashion brand is Uniqlo, is still in EVOLUTION MODE.

- It has got good status in adopting **Detox 2020 plan** and its individual departments has responsibility to adopt proper hazard based screening methodology and recognition of hazardous chemicals out of it. They are also work on clean factory approach, which do not apply on a particular department but to a whole factory to get better results.
- Fast retailing claimed that 98% of its products are **PFC** free till July this year and it requires one more year to achieve 100%.
- Although they have not showcased its data on online. Hence, it needs to work on to be **transparent** by providing data of its suppliers on an ongoing basis and should publish its chemical composition with its wet process suppliers.

6. **G-Star** (Fig. 20)

The position of G-Star occupies in EVOLUTION MODE and is making regular improvement.

- G-Star has on its track to give best for PFC and transparency but somewhere its **Detoxy 2020 plan** has been spoiled that may have bad impact on other parameters in near future. Although, it has better plans to fill those gaps to eliminate hazardous chemicals. To make ensure better results, it needs to create

Fig. 20 Detoxy 2020
plan/PFC
Elimination/Transparency
(http://www.greenpeace.org/
international/en/campaigns/
detox/fashion/detox-catwalk/)

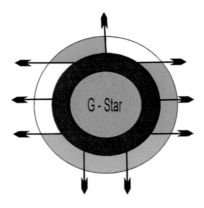

Fig. 21 Detoxy 2020
plan/PFC
Elimination/Transparency
(http://www.greenpeace.org/
international/en/campaigns/
detox/fashion/detox-catwalk/)

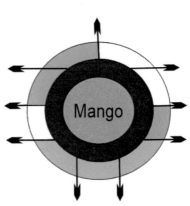

its own MRSL which actively uses a hazard based screening methodology and implements green dimensions.

- This has also been certified to eliminate the use of **PFC** chemicals in 2015 but needs to be regularly updated to further assurance.
- It worked well in **transparency** level by providing all recipes of PFC free chemicals, waste water report investigating report and acknowledging with a list of main vendors or suppliers who provides all manufacturing raw materials.

7. **Mango**

Mango again stays in EVOLUCTION MODE just missing out from AVANT GARDE.

- Mangos detox 2020 plan includes its own individual MRSL, which is regularly updated and implements the fact that there are safe level of hazardous chemicals. It has also a clean factory approach to regulates its targets and apply detox across their whole mill, not only for brand image/products.
- Mango was among the first company to eliminate the use of PFC in its line with its commitment.

Fig. 22 Detoxy 2020
plan/PFC
Elimination/Transparency
(http://www.greenpeace.org/
international/en/campaigns/
detox/fashion/detox-catwalk/)

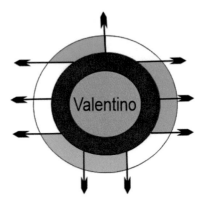

- It has failed in transparency in obtaining the data of hazardous chemicals and its list of suppliers globally online IPE. It failed to publish its suppliers list and composition of natural to man-made ratio that has shown in Fig. 21).

8. **Valentino** (Fig. 22)

Valentino still remains in EVOLUTION MODE with missing modes in all aspects.

- In its **Detox 2020 plan**, This is partly followed due to incomplete communication of its suppliers regarding its priorities, while it has been ban on the use of all PFC chemicals few examples are Chlorinated chemicals which does not come in the list of supplier and products.
- In this investigation we found some **PFC** remain a problem which is exemplary to remove from the products.
- To modify the curve of **transparency**, it must ensure ingoing and outgoing activities of an industry by a clean factory approach that would help in making a good brand image in market. It should be mandatory to publish the pros and cons about product before launch into market with its supplier list including its wet processing stages.

9. **ADDIDAS** (Fig. 23)

Adidas still occupies in EVOLUTION MODE by utilizing hazardous chemicals and selfish approach to make growth faster.

- This brand is trying to be in the list of detoxy plan but from reality test, Physical data has found one step behind to obtaining a green factory approach, pro-active chemicals management and completely knocked out its **Detox 2020 plan**. It needs to be developed **its** own individual MRSL to implement its clean and green factory approach.
- The collected data is all set to target its goals to eliminate **PFC** chemicals from its pocket and maintain its momentum for upcoming years too.
- The curve of **transparency** made it proud, where it has earned lot of money from its customers and claiming it's 50% of global wet process suppliers have disclosed detox data on IPE. To identify the root cause of contaminants and

Fig. 23 Detoxy 2020
plan/PFC
Elimination/Transparency
(http://www.greenpeace.org/
international/en/campaigns/
detox/fashion/detox-catwalk/)

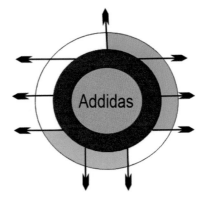

Fig. 24 Detoxy 2020
plan/PFC
Elimination/Transparency
(http://www.greenpeace.org/
international/en/campaigns/
detox/fashion/detox-catwalk/)

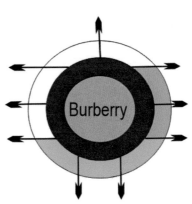

analyse discharge trends, Adidas needs to maintain its momentum as well as publish the reports at the end of cycle.

10. **BURBERRY** (Fig. 24)

Burberry stands still is in EVOLUTION MODE and unable to move beyond this category.

- This brand has come up with its green factory algorithms for its **detox 2020 plan**, Burberry suggests few steps which are strictly communicated to its supply chain and implemented in its MSRL by adding waste water limits. This brand is reflecting its strong company commitments by occupying second world most leader in working with sustainable ways.
- Recently achieved its target to eliminate **PFC** chemicals from its products.
- It has been measured its level of **transparency** that reports around 80% of its global wet processing suppliers have published their data online about their hazardous raw materials to make its image transparent among consumers and documented its whole process from fibre to finishing too.

Fig. 25 Detoxy 2020
plan/PFC
Elimination/Transparency
(http://www.greenpeace.org/
international/en/campaigns/
detox/fashion/detox-catwalk/)

Fig. 26 Detoxy 2020
plan/PFC
Elimination/Transparency
(http://www.greenpeace.org/
international/en/campaigns/
detox/fashion/detox-catwalk/)

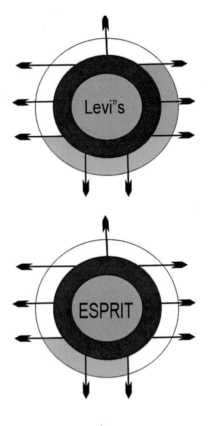

11. **LEVI'S** (Fig. 25)

Levi's still is in EVOLUTION MODE and unable to move beyond this category.

- In this present investigation, levis has shown good commitments with its **detox 2020 plan**, However, levis has adopted scientific approach for sustainable production by following the legacy of making their product more eco-friendly and giving to the birth of water less technology for its denims.
- Recently achieved its target to eliminate **PFC** chemicals from its products.
- It has been measured its level of **transparency** that reports around 80% of its global wet processing suppliers have published their data online about their hazardous raw materials to make its image transparent among consumers and documented its whole process from fibre to finishing too.

12. **ESPRIT** (Fig. 26)

To push back spirit under FAUX PAS category mainly because it has back-trated on its commitment to publish data on hazardous chemical discharges to waste water.

Fig. 27 Detoxy 2020
plan/PFC
Elimination/Transparency
(http://www.greenpeace.org/
international/en/campaigns/
detox/fashion/detox-catwalk/)

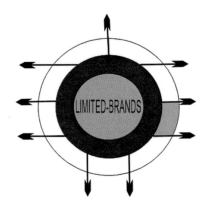

- In esprit **Detox 2020 plan** has found some fundamental flaws that makes it difficult to drive its destination on the path of sustainability. Therefore, it is as important to match the pace of production along with clean image amid consumers by utilizing waste water testing and ETP treatments.
- Eliminating of **PFC chemicals were recorded** in 2014 for the production of spirit goods which it needs to be document in a case study. The workers involved in the manufacturing of spirit goods should take preventive action and to get rid of sustainability risks.
- This above figure shows that there is no longer ensuring that its suppliers will gonna publish detox data, pertaining to its commitments.

13. LIMITED BRANDS (Fig. 27)

Needless to say that Victoria's secret evolved as a big name in lingerie—falls into the faux pas category.

- With its previous records, this brand supported flawed ZDHC chemicals list which resulted it failed to implement its **Detox 2020 plan**. Currently, *Victoria's secret is working closely with ZDHC group for elimination of all hazardous chemicals to make its green image.*
- It has failed to provide new strategy that give positive confirmation that it has eliminated all **PFC** Chemicals and anticipate the key sustainable risks associated with the usage of deadly chemicals in their factories.
- In the pursuit of making more money, the pace of publishing the use of chemicals is not very transparent that come out in their provided discharge analysing report.

14. LI-NING (Fig. 28)

Li-Ning is stuck in faux pas category for failing to improve on its approach to tacking hazardous chemicals in its supply chain, despite some efforts on transparency.

Fig. 28 Detoxy 2020
plan/PFC
Elimination/Transparency
(http://www.greenpeace.org/
international/en/campaigns/
detox/fashion/detox-catwalk/)

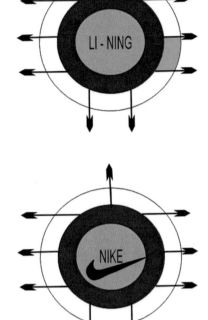

Fig. 29 Detoxy 2020
plan/PFC
Elimination/Transparency
(http://www.greenpeace.org/
international/en/campaigns/
detox/fashion/detox-catwalk/)

- With its previous records, this brand supported flawed ZDHC chemicals list which resulted it failed to implement its **Detox 2020 plan**. Currently, *Li-Ning is working closely with ZDHC group for elimination of all hazardous chemicals to make its green image.*
- It has failed to provide new strategy that give positive confirmation that it has eliminated all **PFC** Chemicals and anticipate the key sustainable risks associated with the usage of deadly chemicals in their factories.
- In the pursuit of making more money, the pace of publishing the use of chemicals is not very transparent that come out in their provided discharge analysing report.

15. **Nike** (Fig. 29)

Nike takes an unfortunate move to faux pas and is the only brand to completely fail on all three of categories accessed.

- On **Detox 2020 plan**, Nike is still struggling on usage of gamut of chemicals for the production of their clothing line and to execute the laws or rules in place for the importing the goods which contain azo dyes that has been ban by several countries. According to report, It does not withstands in implementing its detox commitment instead relying on the inadequate ZDHC chemicals which is missing most of the PFC.

- On **PFC** chemicals elimination, it has succeeded in disposing 90% of the PFC but it still does not commit to eliminate all chemicals from its product it makes.
- On **transparency**, Fashion industry is growing rapidly worldwide and to maintain the pace of production they are not adhere to restricted on one country. Therefore, what they publish are not up to date due to suppressing the inside story of other countries where Nike brand has become paramount for all suppliers and vendors.

7 Conclusion

- After conducting extended research on sustainable design strategies, it can be concluded that the fashion industry involves many constraints for designers. The fashion industry is a very time-constraint industry that has to deal with several seasons and collections per year, and this often results in a high pressure for people working in this industry, including designers. However, the everlasting competition in the fashion industry is crucial in times of economic downturn and this is forcing companies to distinct itself. Due to this fact, companies are starting to recognize the potential of sustainability as a competitive advantage and therefore implement it in their policies. Yet unfortunately some companies 'greenwash' their identity by positive communication on their poor environmental behaviour. Nevertheless, there are many reasons why companies should engage sustainability in their policies and many professionals in the industry believe, according to the conducted interviews, that a change is taking place in the fashion industry and that companies cannot ignore this.
- Recently fashion companies are trying to become more sustainable by organizing sustainable initiatives, but often these are focused on the production (e.g. Levi's water less) or on the consumption of garments (Marks & Spencer) and are not yet initiated at the start of a product lifecycle.
- The design process is often overlooked when creating sustainable initiatives and that is why this research was focused on this aspect. During this research four sustainable design strategies were analysed and reviewed and as a result a handbook for sustainable design was created. When these strategies are implemented by designers it could result in more sustainable products, requiring less water, energy and material use during production and which have a longer lasting lifecycle.
- The question is in how far designers are willing and are able to implement these strategies. Designers are often working under high pressure and it would be a major challenge to implement sustainable strategies unless they are being supported by the company they work for, as it will require extra time and money. Besides, it depends on the interest of every individual designer whether they are

willing to add to a more sustainable fashion industry or not. Nevertheless, still a lot can be improved in an efficient implementation of sustainable design strategies requires a fashion company that has a positive attitude towards sustainability and that is wanting to engage sustainability into their daily procedures.

- Designers cannot implement sustainable design strategies without the support of the company they are working for. The fashion company will need to assist by educating their staff (including designers) and perhaps provide extra money and time for implementation. When these requirements are met, sustainable design strategies could be serving as helpful guidelines for achieving the goal of becoming a more sustainable fashion company.

8 Limitations and Future Scope

- During this research there were several aspects which limited the research on sustainable design strategies. First, the interviews with professionals in the fashion industry were very valuable for my research, however, the input of designers would have been a great addition to the completeness of the research.
- Interviews with designers were very difficult to arrange as they have very little time available and sometimes it can be difficult to get in touch with them. Nevertheless, from the interviews that were conducted I was able to draw more information than I expected on beforehand so this was not a major problem for the outcome of my research. Another limitation for this research was that the strategies could not be properly exanimated within a fashion company.
- The outcome of the implementation of these strategies can therefore only be estimated. The reason for this limitation is that the research was time bound, and that fashion companies might want to implement these strategies on the long term as it is not possible to implement overnight.
- Despite these limitations the research that was completed could be an inspiration and guideline for designers and fashion companies to increase their awareness of the environment in the future.
- During this research many other potential subjects for future research were discovered. A major point of improvement in the lifecycle of a garment is the consumption process of a garment. During the research on sustainable design strategies in fashion, it became evident that the consumption of garments requires most energy and water of all processes in this lifecycle. During the research I came across several interesting opportunities and strategies which could possibly change this process into a more efficient way of consuming garments.

References

Fronteer Strategy & Green Inc., 'The Green Canvas: co-create your sustainability strategy in 5 steps', 2012, <http://www.fronteerstrategy.com/uploads/files/whitepaper/Fs_Whitepaper3-The_Green_Canvas_Co-create_your_sustainability_strategy_in_5_steps-July_2011.pdf> [18 November 2012]

Balakrishnan et al., 'Rewriting the Bases of Capitalism: Reflexive Modernity and Ecological Sustainability as the Foundations of a New Normative Framework', *Journal of Business Ethics* 47 (2003) 4, pp. 299–314

Zizek, S. 'Examined Life' Thoughtware.tv 2010, <http://www.thoughtware.tv/videos/watch/4670-Zizek-we-Should-Become-More-Artificial-> [26 November 2012]

Ehrenfeld, *Sustainability by design: a subversive strategy for transforming our consumer culture*, New Haven (Yale University Press) 2008

Charter and Tischner, 'Sustainable product design' (2001), in: Charter and Tischner (red.), *Sustainable solutions: Developing products and services for the future*, Sheffield (Greenleaf Publishing Limited) 2001, pp. 118–138

Hochswender, Woody. "The Green Movement in the Fashion World." *nytimes*. March 25, 1990

Fletcher, Kate. *Sustainable Fashion & Textiles*. London: Earthscan, 2008

Everman, V., 'How eco is organic cotton?' 2009, <http://life.gaiam.com/article/how-eco-organic-cotton-facts-7-questions> [14–11-2012]

Organic Trade Association, '*Organic Exchange Farm and Fibre Report 2009*', <http://www.ota.com/organic/mt/organic_cotton.html> [14 November 2012]

Fletcher, Kate. *Sustainable Fashion & Textiles*. London: Earthscan, 2008

Nordic Initiative, Clean and Ethical, 'The NICE consumer – Framework for achieving sustainable fashion consumption trough collaboration' 2012, <http://www.bsr.org/reports/nice-consumer-framework.pdf> [20 November 2012]

Carter, K., 'Pandering to the green consumer', *The Guardian*, 13 August 2008

Fletcher, K., 'Fashion and Sustainability', *Sustainable Fashion – Issues to be addressed*, Kolding (Laboratory for Design, Innovation and Sustainability at Kolding School of Design) 2010, pp. 34–41

Black, *Eco Chic: The Fashion Paradox*, London (Black Dog Publishing) 2008

Black, *The Sustainable Fashion Handbook*, London (Thames & Hudson) 2012

Fletcher, Kate. *Sustainable Fashion & Textiles*. London: Earthscan, 2008

Akira, Nakamura. *Fibre Science and Technology*. 2000

Sanfilippo, Damien. *My Sustainable T-Shirt*. 2007

Stolton, Dorothy Myers and Sue. *Organic Cotton From Field To Final Product*. Intermediate Technology Publications, 1999

Greenpeace International (2011). Dirty Laundry: Unravelling the corporate connections to toxic water pollution in China. July 2011 http://www.greenpeace.org/international/en/publications/reports/Dirty-Laundry

Black and Eckert, 'Considerate Design: Empowering fashion designers to think about sustainability' *Sustainable Fashion – Issues to be addressed*, Kolding (Laboratory for Design, Innovation and Sustainability at Kolding School of Design) 2010, pp. 52–65

Fletcher, K., 'Fashion and Sustainability', *Sustainable Fashion – Issues to be addressed*, Kolding (Laboratory for Design, Innovation and Sustainability at Kolding School of Design) 2010, pp. 34–41

Eberle, H. et al., '*Clothing Technology: from fibre to fashion*' Haan-Gruiten (Verlag Europa-Lehrmittel) 2004

Nielsen, Line H., 'The role of the Designer', *Sustainable Fashion – Issues to be addressed*, Kolding (Laboratory for Design, Innovation and Sustainability at Kolding School of Design) 2010, pp. 66–89

Van Nes, N., 'Understanding Replacement Behaviour and Exploring Design Solutions', in: T. Cooper (ed.), Longer Lasting Products: Alternatives to the Throwaway Society, Surrey (Gower Publishing Limited) 2010, pp. 107–132

Allwood et al., 'Well dressed? The present and future sustainability of clothing and textiles in the United Kingdom.', Cambridge (University of Cambridge Institute for Manufacturing) 2006, <http://www.ifm.eng.cam.ac.uk/uploads/Resources/Reports/UK_textiles.pdf> [9 November 2012]

Van Nes, N., 'Understanding Replacement Behaviour and Exploring Design Solutions', in: T. Cooper (ed.), Longer Lasting Products: Alternatives to the Throwaway Society, Surrey (Gower Publishing Limited) 2010, pp. 107–132

LG, 'Caring for Jeans', 2012, <http://www.lg.com/us/laundry/jeans.jsp>

Koo, H., *Design Functions in Transformable Garments for Sustainability.* Dissertation, The University of Minnesota 2012

McDonough and Braungart, *Remaking the Way We Make Things: Cradle to Cradle*, New York (North Point Press) 2002

Blackburn, Sustainable textiles: Lifecycle and environmental impact, New Delhi (Woodhead Publishing Limited)

Gam et al., 'Application of design for disassembly in men's jacket: a study on sustainable apparel design', 8 June 2010 <http://www.emeraldinsight.com/journals.htm/journals.htm?articleid= 1926081&show=html&WT.mc_id=alsoread> [14–10-2012]

Timberland, 'Innovating cadle-to-cradle' 2012. <http://responsibility.timberland.com/product/ cradle-to-cradle-is-coming/> [8 December 2012]

A.R. Horrocks, M. Miraftab and. *Ecotextiles.* woodhead Publishing Limited, 2007

Sanfilippo, Damien. *My Sustainable T-Shirt.* 2007

43. Marchand, A. [et al.], 'Product development and responsible consumption: designing alternatives for sustainable lifestyles', *Journal of cleaner production* 16 (2008), pp 1163–1169

Nielsen, Line H., 'The role of the Designer', *Sustainable Fashion – Issues to be addressed*, Kolding (Laboratory for Design, Innovation and Sustainability at Kolding School of Design) 2010, pp. 66–89

Fletcher, Kate. *Sustainable Fashion & Textiles.* London: Earthscan, 2008

Akira, Nakamura. *Fibre Science and Technology.* 2000

Printed in the United States
By Bookmasters